This volume provides an up-to-date overview of statistical energy analysis and its applications in structural vibration.

Statistical energy analysis (SEA) is a powerful method for predicting and analysing the vibrational behaviour of structures. Its main use is in dealing with structures that can be considered as assemblies of interconnected subsystems which are subject to medium to high frequency vibration sources, i.e. those where the use of deterministic finite element analysis is not appropriate because of the extremely large and complex models that would be required, and the difficulty of interpreting the results of such analyses.

This volume brings together nine articles by experts in SEA from around the world. The opening chapter gives an introduction and overview of the technique describing its key successes, potential and limitations. Following chapters look in more detail at a selection of cases and examples which together illustrate the scope and power of the technique.

This book is based on a *Royal Society Philosophical Transactions* issue under the title 'Statistical Energy Analysis', but an extra chapter by Beshara, Chohan, Keane and Price, discussing nonconservatively coupled systems' is included in this edition.

Statistical energy analysis

An overview, with applications in structural dynamics

Statistical energy analysis

An overview, with applications in structural dynamics

A. J. Keane
University of Oxford

W. G. Price
Southampton University

CAMBRIDGE
UNIVERSITY PRESS

PUBLISHED BY THE PRESS SYNDICATE OF THE UNIVERSITY OF CAMBRIDGE
The Pitt Building, Trumpington Street, Cambridge CB2 1RP, United Kingdom

CAMBRIDGE UNIVERSITY PRESS
The Edinburgh Building, Cambridge CB2 2RU, United Kingdom
40 West 20th Street, New York, NY 10011-4211, USA
10 Stamford Road, Oakleigh, Melbourne 3166, Australia

First published in *Philosophical Transactions of the Royal Society of London,*
series A, volume 346, pages 429–554
Published by The Royal Society 1994
This edition published by Cambridge University Press, Cambridge 1997

Printed in the United Kingdom at the University Press, Cambridge

Typeset in 10/12 point Computer Modern

A catalogue record for this book is available from the British Library

Library of Congress Cataloguing in Publication data applied for

ISBN 0 521 55175 7 hardback

Contents

Preface

This book contains a number of papers dealing with that part of structural dynamics known as Statistical Energy Analysis (SEA). SEA is concerned with vibrations that cannot be readily predicted using deterministic methods as the range of frequencies of interest lies above the first few modes of the structures under study, and where accurate deterministic calculations are no longer useful. As such, SEA is increasingly being used as an adjunct to finite element calculations by engineers dealing with vibration problems.

It was in 1986 that we first began working in this field and at that time, although many papers had been published on the subject only one book had appeared, that by Lyon, which was by then ten years old. Over the last decade research has continued apace and many further publications have appeared. Although Lyon's book has now been revised and republished we felt that work in this area would be served by the production, in a single dedicated volume, of a collection of papers by well respected researchers dealing with the theoretical background to SEA. To that end, in early 1993 we wrote to a number of people who we thought had made a significant contribution to the development of SEA and asked if they would care to submit a paper to such a book. A good number of those asked responded positively and this encouraged us to continue. Most of the resulting papers subsequently appeared as a theme issue in the *Philosophical Transactions of the Royal Society* in March 1994 and are now reproduced here. This volume also contains an additional paper by ourselves and our co-workers that did not appear at that time because of space restrictions in the *Phil. Trans.*

The papers presented are by no means an exhaustive treatise on the theory of SEA but they do give any engineer coming new to the field a good idea of the range of ideas being considered by those researching in the area. The first paper by Frank Fahy deliberately sets out to introduce what follows and also to set the scene on study in this area in a far more comprehensive fashion than can be accomplished in a preface. Subsequent papers then deal with topics as diverse as statistical variations in SEA modelling, modifications to the basic equations of SEA and even parallel techniques which, although using energy as the principal variable of interest, can no longer be strictly called SEA at all. Suffice it to say that we hope this collection of papers will aid and stimulate those whose interest is the vibrations of structures

W. G. Price   Southampton University
A. J. Keane   Oxford University
1996

Statistical energy analysis: a critical overview

By Frank J. Fahy

Institute of Sound and Vibration Research,
University of Southampton, SO9 5NH, U.K.

For the benefit of the 'enquirer within', who may not be familiar with the background and concepts of SEA, this overview opens with a discussion of the rationale for the use of probabilistic energetic models for high-frequency vibration prediction, and introduces the postulate upon which conventional SEA is based. It compares and relates the modal and travelling wave approaches, discusses the strengths and weaknesses of SEA as currently practised and points out needs and directions for future research. Critical discussions of individual contributions to the development of the subject are presented only in as much as they treat specific matters of concept, principle or reliability. The roles of SEA in providing a framework for experimental investigations of the high-frequency dynamic behaviour of systems and in interpreting observations on operating systems, although equally important, are not substantially addressed. Nor are specific experimental techniques which involve considerations of transducers, spatial sampling, signal processing, error analysis and data interpretation, which require a critical review in their own right.

1. Origins

The development of statistical energy analysis (SEA) arose from a need by aerospace engineers in the early 1960s to predict the vibrational response to rocket noise of satellite launch vehicles and their payloads. Although computational methods for predicting vibrational modes of structures were available, the size of the models which could be handled (i.e. the number of degrees of freedom), and the speed of computation, were such as to allow engineers to predict only a few of the lowest order modes of rather idealized models. This posed a serious difficulty because the frequency range of significant response encompassed the natural frequencies of a multitude of higher order modes, the structure could support a number of different wave types, the payload structures were indirectly excited via structural wave transmission, and the transmission paths were circuitous and involved many different forms of structure and contained fluids. It is estimated that the Saturn launch vehicle possessed approximately 500000 natural frequencies in the range 0 to 2000 Hz.

Many previous years of experience with the analysis of sound fields in rooms suggested that the application of deterministic methods of analysis to the prediction of broadband vibrational response of such systems would neither be appropriate nor effective: a concomitant was that it would not be appropriate to estimate the response in terms of local values of kinematic or dynamic variables. Consequently, a form of analysis was introduced in which a system was divided up into subsystems, the subsystem parameters were expressed probabilistically, and the vibrational state of the system was expressed in terms of the time-average total vibrational energy of each of the subsystems, i.e. a global, rather than a local, measure. Vibratory inputs were expressed in terms of time-average input powers, rather than in terms of external forces or displacements. This approach became known as statistical energy analysis.

2. Limitations of deterministic models and analysis

Modern theoretical vibration analysis of complex mechanical structures depends heavily on numerical procedures demanding large, fast computational facilities that can deal with mathematical models representing very detailed idealizations of the physical structures. The computational demands increase with geometric and material complexity, and with increase in analysis frequency. Even today, when computational methods are highly developed and optimized, it is not generally practicable to predict the detailed vibrational behaviour of such structures at frequencies beyond a few hundred Hertz (see, for example, Roozen (1992) who uses 550000 degrees of freedom in a finite element model to study the vibrational behaviour of a 2 m length of aircraft fuselage at frequencies up to 225 Hz).

It is, in principle, possible to extend deterministic computational forms of analysis to higher frequencies, at the expense of rapidly increasing demands in terms of the size of the model and consequent analysis time and cost. However, there is a fundamental physical reason why such extension is ultimately doomed to failure: it is associated with an unavoidable uncertainty about the precise dynamic properties of a complex assemblage of structural components. As frequencies increase, the results of deterministic prediction of frequency response become more and more unreliable. The fundamental reason is that the sensitivity of modal resonance frequencies and relative modal phase response to small changes in structural detail, especially boundary conditions and damping distribution, increases with mode order; and as frequency increases, the responses of almost all systems at any one frequency comprise contributions from an increasing number of modes. The probability distributions of the lower natural frequencies of a simply-supported beam of which the mass per unit length is randomly perturbed are presented by Manohar & Keane (this book). The distributions of the higher order modes begin to overlap, indicating that there is increasing uncertainty in even the order of occurrence of the modal natural frequencies.

There is unavoidable uncertainty about structural detail and material properties associated with manufacturing tolerances and fabrication imperfections, together with environmental and operational influences such as temperature and static load; the high-frequency dynamic properties of joints between components are especially uncertain. As a result, high frequency vibrational responses of individual examples of nominally identical structures are observed to differ, sometimes greatly, as illustrated by figure 1, after Kompella & Bernhard (1993). It is clear that the apparent precision of prediction of response offered by large computational models of complex engineering structures is illusory at frequencies where more than a few modes contribute significantly to the response, and that the high costs of computational procedures based upon deterministic models cannot be justified, particularly since the wealth of detail provided by such procedures is always indigestible and usually unnecessary.

What then is the alternative? It is essentially to treat response prediction from the outset as a probabilistic problem. The high frequency response characteristics of a population of grossly similar systems of which the individual members differ in

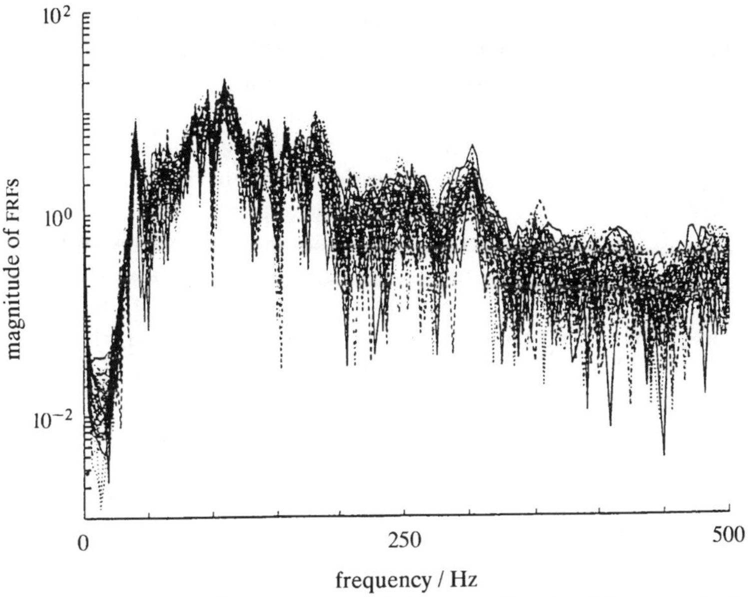

Figure 1. Magnitudes of the 57 structure-borne FRFs for the pickup trucks for the driver microphone.

unpredictable detail may be characterized by their ensemble-average behaviour, together with statistical measures of the distribution of responses about this average. One possible approach to generating an estimate of such a distribution would be to apply a Monte Carlo procedure to a system by randomizing its parameters and properties according to some assumed distributions and repeatedly applying deterministic computational analysis to each member of the set so-generated. A little thought will soon reveal that such an approach is impractical, not only for reasons of cost and time, but because there is no way to model the multi-dimensional joint probability distributions of the large number of parameters and variables.

3. An alternative energy response model

Even though it is impossible to predict precisely the detailed response behaviour of any one physical system to any specific form of high-frequency excitation, it is clear that the gross physical properties such as geometric form and dimensions, together with the material properties, determine the ensemble-average behaviour of a population that shares those properties. It is therefore sensible to seek an approach which utilizes the minimum system description necessary for a prediction of ensemble-average behaviour consistent with the aims of the analyst and with the degree of uncertainty inherent in the prediction on account of imperfect knowledge of the detailed properties of any individual system. Because it is unrealistic, and generally unnecessary, to attempt to predict the response at every point of a system, it is sensible to seek a global description of the response of individual components of the system within which the ensemble-average response is expected to be reasonably uniform: this is a necessary condition for the usefulness of a global response prediction, since it would not be sensible to attribute a single response level to contiguous systems which are likely to exhibit greatly different levels.

These considerations lead to the adoption of a gross parameter model. This is conceptually divided up into subsystems that either have significantly different gross properties, or are separated from contiguous subsystems by structural elements, which form significant barriers to transmission of vibration from the source(s) of excitation. Because the relative amplitude and phase of the frequency response at individual points within any subsystem are essentially unpredictable, it is reasonable to treat these response quantities as being random, thereby introducing the concept of treating the frequency response functions at individual points as a population of which only statistical descriptors such as the ensemble mean value and measures of the spatial variation about this mean are meaningful.

The total time-averaged energy of vibration is an attractive global descriptor of subsystem response since it is phase-independent and is subject to the fundamental constraint imposed by the principle of conservation of energy. The multi-degree-of-freedom model or finite element model of a subsystem is replaced by a single degree of freedom, and the conventional frequency response function is replaced by a subsystem energy response function. Naturally there is a penalty; no information is available about spatial distributions of response variables such as strain or acceleration: however, these can also be represented by probabilistic models (see, for example, Stearn 1970). Once this energy response descriptor is adopted, the description of subsystem interaction in terms of rate of energy exchange (or 'power flow') and the description of subsystem excitation in terms of power input from external sources naturally follow.

In summary, the energy response model comprises a set of subsystems described by their gross geometric forms and dynamic material properties, which are subject to external power inputs, and which receive, dissipate and transmit vibrational energy. The output of an energy response analysis is an estimate of the equilibrium time-average energies of vibration of the various subsystems.

4. Statistical energy analysis

(a) *Frequency-average estimates*

Statistical energy analysis (SEA) in its current state of development provides a means for estimating the equilibrium energies of a network of subsystems which are subject to an assumed distribution of external sources of time-stationary vibrational power input. For historical and practical reasons, it does not deal directly with ensemble-mean energy response functions for a population of systems that share the same gross parameters, but which differ in detail, as discussed above. Instead, it is used to estimate the frequency-average value of the energy response functions of individual archetypal subsystems over intervals of frequency that are adjudged large enough to justify simplifying assumptions concerning the coefficients which relate time-averaged vibrational power flow between subsystems to their equilibrium energies.

The relation between frequency-average and ensemble-average energy response is the subject of current research and is not yet fully resolved. However, the use of frequency-average energy response is more practical for the purposes of experimental SEA, which plays a vital complementary role to predictive SEA. In many cases of engineering interest, the power transmission and dissipation coefficients associated with selected subsystems are not computable, and consequently they must be derived from experiments on physical systems. It is not generally practicable to measure ensemble-average coefficients on populations of physical systems, but it is simple to implement frequency averaging of experimental response data obtained from one system. It is worth noting here that SEA may also be used to interpret response data collected from operating systems for the purposes of selecting appropriate vibration and noise control measures. It has also been used in an inverse mode to infer the locations and magnitudes of sources of external power input from experimental estimates of subsystem energies and a previously established SEA model.

The evaluation of the coefficients that relate power transfer between subsystems to their equilibrium energies lies at the heart of the engineering application of SEA. There are three main approaches to this problem. In the *modal* approach, the equations governing the dynamic behaviour of the subsystems are expressed explicitly in terms of expansions of the vibrational fields in series representing the uncoupled natural modes of the subsystems: the choice of boundary conditions which express the decoupling is a matter of analytical simplicity and physical characteristics (Maidanik 1976). It is then possible, in principle, to express the multi-mode power transfer coefficient as an estimate of the average of the individual mode-pair coupling coefficients, subject to certain assumptions regarding modal frequency distributions and coupling strengths using the results of coupled oscillator analysis. This approach is rather well suited to vibroacoustic problems involving acoustic interaction between enclosed volumes, gases and solid structures, because coupling is normally weak, the *in vacuo* modes of the structure can be used, and the statistical properties of rigidly bounded fluid volumes have been exhaustively researched (see, for example, Fahy 1970).

The explicit modal approach is not ideally suited to coupling between solid structures, especially where they are not uniform and isotropic. Consequently a travelling *wave* approach is used, in which the vibrational fields are modelled as superpositions of travelling waves (usually plane), and transfer between subsystems is evaluated from considerations of wave transmission and reflection at subsystem interfaces. The wave intensities distribution is related to subsystem energy density, to the dispersion relation of the waves and to the subsystem geometry. Wave power transmission coefficients can be evaluated using a combination of statistical models of incident, reflected and transmitted wave fields with relatively small-scale deterministic models of interface dynamics analysed by computational procedures such as finite element analysis. A modern development of the wave approach is presented by Langley & Bercin (this book).

The *mobility* approach utilizes the concept of dynamic mobility, or impedance (point, line or wave) to express the result of interaction between coupled subsystems. To deal with probabilistic models of the subsystems, averages of various forms (spatial, frequency, ensemble) are applied to these dynamic characterizations of subsystems. This approach is explained by Manning (this book) and evaluated by comparison with experiment by Cacciolati & Guyader (this book).

(*b*) *The modal approach to* SEA

Where two subsystems interact to exchange energy, it is possible to express the interaction in terms of the modes of the uncoupled subsystems, of which selection of the appropriate boundary conditions depends upon the physical natures of the subsystems. SEA has traditionally been developed in terms of this modal interaction model, and the SEA equations have been developed as an extension of the relationship between the energies of vibration of two conservatively coupled, viscously damped oscillators subjected to white noise excitation, and the rate of exchange of energy, or power flow. This exact relation is (Scharton & Lyon 1967)

$$P_{12} = g[E_1 - E_2], \tag{1}$$

where P_{12} is the time-average power flow, E_1 and E_2 are the time-average equilibrium energies of the oscillators when subject to the given excitation but with the motion of the other oscillator 'blocked', and g is a power transfer coefficient. (In this simple case, the relation also holds for the actual coupled energies, but, of course, g is different.) The coefficient g is a function of the properties of the coupling elements, the oscillator half power bandwidths (damping), and the natural frequencies of the uncoupled oscillators: it is particularly sensitive to the difference of natural frequencies of the blocked oscillators.

Formal extension of this analysis to power exchange between two sets of oscillators is analytically straightforward, but the resultant expressions for power exchange involve terms that express correlations between the motions of all the oscillators and is not only algebraically unwieldy, but useless for practical purposes. If it is assumed that the motion of each oscillator within one set is uncorrelated with the motion of every other within the same set, these terms disappear and the total set-to-set power exchange can be expressed as the sum of the power exchanges between individual pairs of oscillators, one from each set.

Since the motion of an individual oscillator is determined by the combined effects of any external excitation and the reaction forces produced by interaction with other oscillators, neglect of intra-set correlation requires two assumptions: (i) that the external forces on each oscillator are uncorrelated; (ii) that coupling is sufficiently 'weak' to ensure that indirect intraset oscillator interaction via any one oscillator of the other set is negligible compared with direct inter-set oscillator interaction. Extension of this weak coupling model to energy exchange between the modes of distributed elastic systems is based upon an assumption of uncorrelated modal generalized forces and weak coupling between the two subsystems. Uncorrelated modal forces can only be generated by delta-correlated or 'rain-on-the-roof' force fields: the basic SEA relation does not hold in the case of locally concentrated excitation, except in the sense of an ensemble-average over all locations. Further, to use the concept of a 'modal-average' coupling coefficient, it is assumed that energy is shared equally by modes resonant within the band of analysis (equipartition of modal energy). Under these (rather restrictive) conditions, the two-oscillator result may be extended to an analogous relation between the power flow between the two sets of oscillators which represent the uncoupled modes of the subsystems, and the average stored energy per oscillator (modal energy) of each subsystem, thus (Scharton & Lyon 1967);

$$P_{12} = M_{12}[E_1/n_1 - E_2/n_2], \tag{2}$$

in which P_{12} is the net power flow between the subsystems, E represents total subsystem energy, n represents modal density, which is the inverse of the local average frequency spacing between successive modal natural frequencies and M_{12} is a modal-average 'power transfer coefficient'. Newland (1969) derives an equivalent relation for weakly coupled oscillator sets on the basis of a small parameter, perturbation approach and Zeman & Bogdanoff (1969) do the same for extended elastic systems. On the basis of a criterion for 'weak' coupling which is closely related to that adopted by Newland, namely that the Green functions of the subsystems are little affected by coupling, Langley (1989) also confirms the general relation expressed by (2). Gersch (1968) generalizes the results of Scharton and Lyon and of Newland to cases involving non-conservative coupling, giving exact expressions for a three-oscillator case. The term 'power transfer coefficient' used herein is not standard; it is conventional in SEA to express (2) in terms of a quantity known as the 'coupling loss factor' which is defined by analogy with the dissipation loss factor thus:

$$P_{12} = \eta_{12}\,\omega E_1 - \eta_{21}\,\omega E_2. \tag{3}$$

It transpires that the coupling loss factor is a function of the spatial extent of a subsystem, and that the product of coupling loss factor and modal density is a physically more significant quantity which is both independent of subsystem extent and independent of the direction of power flow considered.

Comparison of (2) and (3) indicates that the power transfer coefficient M_{12} is equivalent to $\eta_{12}\,\omega n_1 = \eta_{21}\,\omega n_2$. It represents a form of mode-pair average of the coupling coefficient g, and is formally a function of the physical coupling between the subsystems, the degree of spatial matching of the mode pair shapes integrated over the common interface, the modal dampings and the relative distributions of mode pair natural frequencies. M_{12} is independent of modal damping only in the case of weak coupling, and only then if a certain form of probability distribution of mode pair frequency difference obtains. A condition of 'strong' coupling, in SEA terms, produces very similar values of average modal energy in all strongly coupled subsystems, irrespective of the distribution of input power. It will also necessarily produce correlated modal responses and non-uniform modal energies in subsystems which are not subject to external forces, but are driven only through couplings to other subsystems. It may be argued that the condition of 'weak coupling', necessarily requires that $M_{12} \ll M_1$ and M_2 for all pairs of connected subsystems (Smith 1979). Keane & Price (1987) present an analysis of power flow between multi-mode systems strongly coupled at a single point in which subsystem receptances are used to derive an exact analytic expression for the power transfer coefficient. Simplification of the exact expression is achieved by ensemble averaging and using a Monte Carlo approach to evaluating the behaviour of various terms in the expression. They demonstrate that the basic power flow-energy difference relation holds for all strengths of coupling, but that the expressions for power transfer coefficient which obtain under conditions of weak coupling must be modified to account for strong coupling, in agreement with Lyon (1975) and Mace (1992). The transition from weak to strong coupling in a system of spring-coupled rods is clearly illustrated by the work of Keane & Price (1991), in which the power flow increases monotonically with increase of spring stiffness up to the transition point, after which it becomes independent of coupling strength.

Except in the cases of highly idealized mathematical models, explicit averaging of mode pair coupling coefficients is not possible in practical systems for the reasons discussed above; modal parameters are normally only describable in a probalistic sense. The dependence of the power transfer coefficient on the probability distribution of mode pair natural frequency differences suggests that it will be rather sensitive to perturbations of the subsystem physical parameters under conditions where the frequencies are well separated in terms of typical modal bandwidths, i.e. low modal overlap. In this respect, modal natural frequency distribution statistics for practical systems have been inadequately studied.

The relationship between power flow and modal energy difference expressed by (2) forms the central 'postulate' upon which the edifice of SEA has been constructed.

It is analogous to the equation of conductive heat flow in which the average modal energy is analogous to temperature.

One implication of (2) for practical applications is that SEA cannot be expected to give accurate estimates in cases where the contiguous subsystem modal energies are very similar (close to equipartition of modal energy) since given fractional errors in the estimate of either or both will give rise to much larger large fractional errors in the difference, and hence in the power flow estimate. As shown by Heckl & Lewit (this book), this is especially relevant where SEA power balance equations are used, via loss factor matrix inversion, to estimate coupling loss factors from measured subsystem responses. This requirement has implications for the selection of subsystem definitions and boundaries, and rules out the use of SEA for systems through which vibrational waves travel with so little attenuation or reflection that modal energies are all very similar. On the other hand, excessive damping renders inappropriate the concept of modal energy as a global descriptor of vibrational state.

The modal (oscillator) interaction model is reasonably satisfactory as a conceptual basis for SEA but it is generally not well suited to computation in structural engineering applications and it raises difficulties of physical understanding of the power exchange process since uncoupled modes are generally chosen to satisfy boundary conditions which allow no energy transmission across the boundary. Consequently, in the practice of SEA, the modal model is rarely used directly to predict structural vibration, although numerous SEA validation exercises have been performed using the known modal characteristics of uniform subsystems such as rods, beams, rectangular plates and circular cylindrical shells in various combinations (see, for example, Keane & Price 1987; Davies 1981; Dimitriadis & Pierce 1988). The modal approach is, however, more practical in cases of vibroacoustic interaction between extended structures and volumes of contiguous fluid, in which power transfer takes place over the surface of the common boundary, and the coupling loss factor is influenced by the structural boundary conditions. In practice, the energy distributions and the processes of subsystem interaction are more commonly expressed in terms of propagating wave models.

Woodhouse (1981) presents a treatment of vibration transmission between coupled multimode systems which is based upon the classical multi-degree-of-freedom generalized coordinate representation and which facilitates the introduction of extra degrees of freedom as arbitrary coupling elements. This paper, together with the thought provoking overview of high-frequency structural vibration of Hodges & Woodhouse (1986), is recommended to the reader who wishes to acquire greater physical insight into the foundations of SEA.

(c) *The wave approach to SEA*

In the wave approach to SEA, the vibrational fields of subsystems are represented in terms of superpositions of travelling waves, and the power transfer coefficients between subsystems are evaluated from consideration of wave reflection and transmission at their junctions. In the case of spatially uniform junctions and subsystems, the harmonic plane wave (single wavenumber) transmission coefficients are simple functions of the input wave impedances (or mobilities) of the uncoupled subsystems evaluated at the joint. It should be note carefully that, as explained by Manning (this book), the mobility of a spatially extended joint is not equal to the sum of the mobilities of the connected subsystems.

Junction mobilities are conventionally evaluated on the assumption that transmitted waves returning to the junction after reflection from other boundaries are uncorrelated with the directly transmitted waves. This assumption is clearly unjustified where systems exhibit distinct resonances, because resonances and modes are the result of interference between coherent waves travelling in different directions: hence, power transfer coefficients for systems of low modal overlap are likely to be rather sensitive to perturbations of subsystem parameters.

Just as mode shapes, dampings and frequencies are described in a statistical sense in the modal approach, so the distribution of amplitudes, phases, attenuation and directions of wave propagation must be represented by statistical distributions in the wave approach. For example, the wave equivalent of the uncorrelated modal response assumption is an assumption of zero correlation between the set of travelling plane waves by which a subsystem vibration field is modelled: the equivalent of equipartition of subsystem modal energy is a smooth, sometimes uniform (diffuse), angular distribution of wave intensity. It may be shown that wave intensity may be directly related to modal energy, which relation provides the link between the two approaches. A profound implication of the assumption of uncorrelated travelling wave components is that reactive wave intensity (non-propagating energy density) does not exist. This is in stark contrast to the modal model, because modes involve predominantly reactive intensity. A related distinction is that, unlike the modal coupling model, in which coupling coefficients are, in general, functions of subsystem damping, the power transfer coefficients derived from wave transmission models of coupled semi-infinite subsystems are independent of damping. Given these gross differences in the assumed energetic characteristics of the modal and wave models, it is somewhat surprising that they produce compatible predictions.

There is a fundamental distinction between the modal and wave representation. Uncoupled subsystem modes, although unpredictable in a deterministic sense, are entities whose characteristics depend inherently upon the assumed geometric and dynamic boundary conditions, and which are associated with particular natural frequencies. Propagating waves are controlled only by the dynamic properties of the wave-bearing medium and are essentially independent of boundary conditions: the latter only influence wave scattering at the boundary. Consequently, the travelling wave representation involves no consideration of natural frequencies or assumptions about their distribution. In its favour, it raises fewer problems with regard to the representation of the action of distributed dissipative mechanisms acting within the body of a subsystem, and at boundaries, than the modal approach, where the question of the representation in terms of real or complex modes is a vexed one. These distinctions make the wave approach more attractive in terms of qualitative appreciation of the physical behaviour of a complex system. In fact, the most commonly applied method of evaluating the coupling loss factor in (3) is based upon wavefield modelling, and not modal modelling.

However, although the travelling wave approach is more appropriate at frequencies where numerous modes make significant contributions to the response at any one frequency (i.e. the modal overlap factor greatly exceeds unity) it may prove to be less helpful in developing means of making confidence estimates for SEA predictions in frequency ranges where distinct individual resonance peaks are observable in the energy-response functions, since a probabilistic wave model excludes the phenomenon of resonance. When the modal overlap factor is less than unity, a probabilistic model of the occurrence of modal resonance frequencies will be a necessary ingredient of the analysis of variance of response estimates. (It should be noted that the subsystem impedances which control the wave power transmission coefficient will exhibit strong frequency dependence under conditions of low modal overlap (because of coherent (modal) interference), and statistical models of the distribution of impedance components may also be used to study uncertainty in SEA response estimates.)

(*d*) *The mobility approach to SEA*

The dynamic properties of system may be characterized by means of a transfer function which relates a harmonic input force or displacement field to the resulting response. In vibration analysis, these transfer functions take two principal forms: impedance relates force input to velocity response, and mobility is the corresponding inverse; where both quantities are specified at the same point or line, they are known as 'direct' or 'input' impedance and mobility. The most common forms correspond to 'point', 'uniform line' and 'plane wave' inputs. When a vibrational wave in a subsystem impinges upon an interface between it and a subsystem having different dynamic properties, wave reflection and transmission occur, which may or may not involve non-specular scattering, diffraction and refraction. The resulting forces and displacements at the interface depend upon the form of the incident field and the dynamic properties of both systems. The power transmitted through the interface is given by the time-average product of the associated forces and velocities. These are related by the interface mobility (or impedance) appropriate to the form of incident wave. Hence, wave power transmission coefficients can be expressed in terms of mobilities. Because the incident wave power can be related to the modal energy of the associated system as explained in the previous section, the power transfer coefficient of (2) can be evaluated in terms of the junction mobilities. These, in turn may be determined in a frequency or ensemble average sense for specific physical forms of subsystem, together with assumptions concerning the statistical distributions of the subsystem parameters. As explained by Manning (this book), an average over an exact power transfer expression is approximated by inserting average mobilities into the expression. This constitutes the essential approximation at the heart of the mobility approach to SEA.

(*e*) *Modal energy and wave intensity*

As seen from (2), the fundamental SEA relation between power flow and modal energy difference is analogous to a heat flow equation for bodies at different temperatures. Modal energy may be thought of as a measure of subsystem temperature. It is not immediately obvious how this analogy may be explained in terms of the physics of vibrational energy transfer between subsystems. An examination of analytical expressions for the modal energies of simple, uniform, isotropic subsystems provides some elucidation of this issue. (Non-isotropic or curved shell structures require special analysis because they do not possess direction-independent wave group speeds and therefore do not support diffuse fields. The wave intensity distribution analysis of Langley & Bercin (this book) is relevant to such cases.) The average density of modal natural frequencies associated with any vibrational wave-type is directly proportional to the spatial extent of the bounded medium (length, area or volume) and inversely proportional to the wave group speed (energy propagation speed). For example, the modal density $n(\omega)$ of a uniform one-dimensional system is given by $2L/c_g$, of a uniform, isotropic two dimensional plane structure of surface area S is given by $kS/\pi c_g$ and of a uniform, isotropic volume V

is given by $k^2V/2\pi^2 c_g$. Hence the modal energy E/n is seen to be proportional to the product of energy density (energy per unit length, area or volume) and group velocity, which has the dimensions of wave intensity. This relation reveals the link between the modal and wave models. It should be noted that equipartition of modal energy only corresponds to an idea diffuse wave field if many modes are excited to similar energy levels and wave group velocity is independent of direction.

Because conventional SEA is only applicable to subsystems that are rather reverberant (so that the energy density within each is rather uniform), only some proportion of the total energy density is associated with energy transport towards an interface with a contiguous subsystem. Of course the assumption of a completely diffuse field is inconsistent with net transport of energy in any one direction, since the intensity of waves partly reflected from an absorbing boundary must be less than those approaching the boundary. However, provided that coupling is weak, in the sense that the wave power transmission coefficients are much less than unity, it is permissible, but not necessarily valid, to make the diffuse field assumption. It is therefore possible analytically to relate the energy density to the vibrational power incident upon a subsystem boundary. For example, modal energy of a one-dimensional system is given by $E c_g/2L$, which, if the energy density is assumed to be uniformly distributed (one-dimensional 'diffuse' field assumption) corresponds to the assumption that half the energy density is transported in one direction and half in the other (this is clearly only reasonable for reverberant, weakly coupled subsystems).

The wave intensity analysis of Langley & Bercin (this book) represents the angular distribution of wave intensity in a uniform plate or curved shell structure by means of a Fourier series expansion in orders of angular coordinate, and thereby provides a means of representing non-diffuse fields: the intensity distribution is assumed to be homogeneous throughout a uniform subsystem. The power incident upon any the subsystem boundary may therefore be explicitly determined from the geometry of the boundary. SEA-type power balance equations may be written for each Fourier component. Wave diffraction at, for example, interfaces that are small in comparison with a structural wavelength remains to be incorporated in this model.

5. General modes of application of SEA

A power balance equation in the form of (2) may be written for each subsystem, and the set is expressed in matrix form as

$$\begin{bmatrix} M_1 + \sum\limits_{i \neq 1}^{k} M_{1i} & -M_{12} & \cdots & -M_{1k} \\ -M_{12} & M_2 + \sum\limits_{i \neq 2}^{k} M_{2i} & \cdots & -M_{2k} \\ \vdots & & & \\ -M_{1k} & \cdots & \cdots & M_k + \sum\limits_{i \neq k}^{k} M_{ki} \end{bmatrix} \begin{bmatrix} E_1/n_1 \\ E_2/n_2 \\ \vdots \\ E_k/n_k \end{bmatrix} = \begin{bmatrix} P_1 \\ P_2 \\ \vdots \\ P_k \end{bmatrix}$$

This set of equations may be used in various modes. One is the 'response prediction' mode, in which the power transfer coefficients and input power distributions are known, or assumed, *a priori*, and the resulting energy states of the subsystems are derived by inverting the matrix of power transfer coefficients. In another mode, known input powers are applied to the individual subsystems in turn, and experimental estimates are made of the resulting total energies of all the subsystems, so that estimates can be made of the dissipation loss factors of each and the coupling loss factors between each. A major assumption underlying the principle of this method is that the power transfer coefficients are invariant with respect to the distribution of external input power to the subsystems; this is unlikely to be the case in practice. (The power transfer coefficients matrix is not determined directly because it requires knowledge of the subsystem modal densities.) In a third mode, experimental estimates are made of the relative modal energies in operational systems, so that directions of energy flow can be inferred for the purposes of identifying principal power transmission paths in multiply connected networks of subsystems. A study of the effectiveness and problems of these modes of application is presented by Heckl & Lewit (this book). In the 'source identification' mode, operational energy distributions are combined with assumed, or pre-determined, loss factors to quantify the power inputs into each subsystem. This is perhaps the most problematic mode of application because, in principle, it admits of an infinite number of solutions, and its success depends upon the physical validity of the assumed model of uncorrelated source distribution, which, in turn, depends upon the analyst's understanding of the physical excitation mechanisms and processes involved.

Each experimental model requires estimates to be made of the uncertainties in the experimentally determined estimates, so that the associated uncertainties in the derived quantities may be estimated. This procedure involves, *inter alia*, estimates of spatial sampling errors, modal masses, spectral estimation errors, numerical matrix inversion errors, and deserves a critical review in its own right.

6. General advantages of SEA

(*a*) It requires only a relative coarse idealization of the physical system that possesses relatively few gross parameters.

(*b*) It uses a small number of degrees of freedom per subsystem.

(*c*) The use of an SEA model allows an analyst to retain a 'feel' for the behaviour of the system in terms of the physical influence of system parameters. Because the latter are few in number, and computer runs are very cheap and quick, SEA is very useful in early design exercises.

(*d*) It forms an excellent framework for designing and performing experiments for the purpose of determining power transfer and loss coefficients, and hence for providing a designer with a model on which to base strategies for modifying system response.

(*e*) SEA is based upon the principle of conservation of energy, and violations of this principle can easily be detected. Also, an assumption of equipartition of modal energy between a directly-excited and a non-directly-excited subsystem provides an estimate of the response of the latter with prior knowledge of only the order of magnitude of the transfer coefficient and damping. Alternatively it provides a conservative estimate of the upper limit of response.

(*f*) An SEA model, once constructed in the laboratory, can, in principle, be used to detect and quantify sources of external power input to systems on the basis of measured operational response data.

(*g*) Experimental estimates of subsystem modal energies indicate in which direction power is flowing.

7. Current deficiencies of SEA

(*a*) *Predictive confidence*

SEA is inherently probabilistic and yet there is no established procedure for making estimates of confidence in the predicted results, although some rather speculative expressions have been presented (Lyon 1975). Other contributions include those of Keane & Price (1987), Mace (1992), Fahy & Mohammed (1992) and Craik *et al.* (1991). Designers therefore cannot currently rely upon the results of parametric sensitivity studies, or satisfy design requirements at any specific level of confidence. In particular, the influence of modal overlap factor and number of modal resonance frequencies in the analysis band on confidence intervals has not been established. Until some formal procedures for this purpose are developed, SEA will be relegated to the 'high frequency' league, although the modal SEA model is valid in any frequency range. Theoretical and experimental research is currently in progress to establish ensemble energy response and power flow statistics for assemblages of simple subsystems but much work needs to be done before any generally applicable procedure can be developed. One example of the products of this research is represented by Manohar & Keane (this book), which revives a previously proposed measure of uncertainty of the properties of a subsystem in the form of a 'statistical overlap factor'.

(*b*) *Tonal or narrow band excitation*

SEA cannot deal with narrow band or tonal excitation without supplementary statistical data for energy response functions for directly driven and indirectly driven subsystems of various generic forms, which is not readily available. There exist many problems of tonal excitation of systems at frequencies many times their fundamental natural frequency, including excitation of rocket components by high-speed pump noise, noise generation in helicopters by gear box vibrations, and high speed compressor vibration. A reasonable approach to this problem would be to estimate the response as if generated by a band-limited white noise source using SEA, and then to use statistical estimates of the likely distribution of single frequency energy response about this estimate. The current work on uncertainty of SEA predictions, referred to above, will yield data of assistance in this respect.

(*c*) *Spatial distributions of response within a subsystem*

SEA produces estimates of global energy in a subsystem and gives no information about the spatial distributions of the field variables within each subsystem. Systems fail, or malfunction, because of the occurrence of excessive local response, not global response. Supplementary procedures and data are necessary for the prediction of local response. Examples can be found in Stearn (1970) and Norton & Fahy (1988).

(*d*) *Periodic systems*

Conventional SEA gives extremely unconservative estimates of the transmission of vibration along chains of similar systems (periodic systems) (see, for example, Blakemore *et al.* 1992): errors of 30 dB or more have been observed. Few 'real' systems are perfectly periodic, but it is not known how this error varies with the degree of disorder of almost periodic systems. One explanation of this behaviour is that even if a diffuse field is generated in a directly excited subsystem, the subsystem interface will spatially filter the wave field incident upon it, so that the transmitted wavefield will be especially rich in those waves which pass most easily through the interface, thereby rendering the field less and less diffuse as it passes through successive interfaces from subsystem to subsystem, and allowing progressively easier transmission through each. The wave intensity approach of Langley & Bercin (this book) overcomes this failing of conventional SEA by separating the subsystem fields into diffuse and non-diffuse components. It also yields estimates for coupling loss factors between non-contiguous subsystems. Keane & Price (1989*a*, *b*) consider the effect on power flow behaviour of the non-random distribution of natural frequencies in ordered systems, and deviations from the results of frequency-averaged SEA analysis of large systematic variations in modal density associated with periodic systems: they also address the effect of small variations from pure periodicity.

(*e*) *Indirect coupling*

Comparison with the results of alternative forms of analysis, and experimental observations, indicate that indirect coupling must often be included if an SEA model is used, especially where not all subsystems are subject to external excitation, and where coupling is not weak (Keane & Price 1992; Langley & Bercin, this book). This does not mean that energy can pass from one subsystem to another non-contiguous subsystem without passing through intermediate subsystems; it simply means that the conventional SEA model does not correctly represent all processes of energy transport. Heron (this book) introduces the concept of wave energy ‘tunnelling’ in which the damping of intermediate subsystems influences the indirect power transmission coefficients.

(*f*) *Systematic deviation of power transfer coefficients from those based upon semi-infinite subsystem wave power transmission coefficients*

There is theoretical and experimental evidence to suggest that, under conditions of low modal overlap, and strong coupling, power transfer coefficients systematically deviates from semiinfinite subsystem wave power transmission values (see, for example, Mace 1992; Heckl & Lewit, this book; Cacciolati & Guyader, this book; Gibbs 1976; Fahy & Mohammed 1992; Boisson *et al.* 1985). This is not surprising, because the wave transmission approach produces power transfer coefficients independent of natural frequency distribution and subsystem damping. This deviation, if it occurs, would create significant bias errors in SEA estimates, made on the basis of wave transmission theory. It may well be related to the observation of Scharton (1971) that the power injected by finite impedance sources into multi-mode media under conditions of low modal overlap fall below the power injected by the same source into a non-reactive (freefield) load. As far as one subsystem is concerned, a contiguous subsystem is a source of vibrational power injection possessing a finite, frequency-dependent source impedance.

(*g*) *Highly damped subsystems*

The SEA model assumes that subsystems are quite reverberant so that the energy density in a subsystem is reasonably uniformly distributed. This is not so in cases where waves are rather highly damped as, for example structural waves in aircraft fuselages and sound waves in air cavities containing sound absorbent material.

(*h*) *Coupling by concentrated elements*

In many systems of practical interest, extended subsystems are connected by concentrated elements which may exhibit internal resonant behaviour, for example hull frames on ships. The wave transmission coefficient of such structures becomes highly frequency-dependent close to the associated resonance and anti-resonance frequencies,which are rather sparse in frequency, so that a subsystem receiving energy via such an element from an adjacent subsystem which is itself subject to broadband excitation is not effectively subject to broad band excitation. This general problem is addressed by Woodhouse (1981), and the special case of symmetric resonant couplings is investigated by Allwright *et al.* (this book), who conclude that such an element may either be treated as part of an interface between two subsystems, for which a diffuse field power transmission coefficient may be determined by normal classical elasto-dynamic analysis, or treated as a separate subsystem in an SEA sense.

(*i*) *Non-conservative couplings*

In many practical systems, a major part of the energy dissipation occurs in the joints and connections between component structures. Studies have been made of power transfer in coupled oscillators by Lyon & Maidanik (1962) and Fahy & Yao (1987) that indicate that the oscillator energy sums, as well as differences, enter into the power-flow–energy relation. Gersch (1968) formally analyses power flow between systems of non-conservatively coupled oscillators, but the results are not framed in a manner amenable to interpretation in an SEA framework. It is not known to what degree a concentration of energy dissipation in the interface region affects the accuracy of SEA predictions. Non-conservative coupling is examined in some detail in the final chapter of this book (which was not in the previous publication).

8. How successful is SEA in practice?

It is difficult to evaluate the true degree of success of SEA as an analytical technique in predicting the behaviour of practical engineering structures because, in general, only the results of successful applications are published and few authors present sufficiently critical and rigorous analysis of cases where it has been unreliable. However, it has been conspicuously successful in dealing with vibro-acoustic problems involving the interaction of broadband sound fields in air with structures such as space satellites and launch vehicles, and has also given reliable estimates of structure-borne sound transmission in buildings (Craik 1991) and ships (Plunt 1960). It has also been successfully used to provide a means for empirical studies of structure-borne sound transmission in road vehicles (Lalor 1991). It must be emphasized that these citations are selected principally because they represent examples of proven effectiveness with which I am most familiar: no doubt many others of similar merit exist in the technical reports of laboratories worldwide.

It is, perhaps, a measure of the value of the SEA (warts and all) to the engineering community that computer software packages for its application are now commercially available and finding a welcome as a tool for handling otherwise intractable problems. A major benefit of the increasing use of such packages is that the inherent weaknesses or failings of SEA as currently practised will become more widely appreciated and more expeditiously remedied, a process which may be aided by research into the topics suggested in the following section.

9. Areas for future research

The following topics are considered to merit further study and investigation.

(i) The influence of subsystem network topology and geometric form on the ensemble statistics of power flow and energy response functions.

(ii) Development of a generalized method for predicting indirect power transfer coefficients and further experimental verification studies of the phenomenon.

(iii) The effects of degree of disorder on power flow in nearly periodic, plate and shell systems.

(iv) The representation of dissipative couplings in SEA-type models. (See also the final chapter of this book.)

(v) Modelling and analysis of power flow between structures coupled by a dense fluid medium, including free surface effects.

(vi) Natural frequency distribution statistics for practical structures under conditions of low modal overlap.

(vii) The influence of strength of coupling, modal overlap and mode count (bandwidth) on frequency-average power transfer coefficients. Deviation from semi-infinite wave power transmission coefficients under conditions of low modal overlap and strong coupling.

(viii) Relation between ensemble and frequency band energy response statistics.

(ix) Experimental studies of vibrational intensity distributions in structures for comparison with the predictions of wave intensity analysis.

(x) Development of reliable means for theoretically handling highly non-uniform structural components.

(xi) Experimental studies of spatial distribution statistics of response variables for practical structures.

(xii) Development of an SEA-equivalent approach to the prediction of response of complex systems to transient inputs.

The monograph by Lyon (1975) remains an inspirational source of concepts and analytical strategies in the field of probabilistic analysis of high-frequency vibration, and it will repay even the hardened SEA veteran to revisit this seminal publication from time to time.

References

Blakemore, M., Myers, R. J. M. & Woodhouse, J. 1992 The statistical energy analysis of cylindrical structures. In *Proc. Second Int. Conf. on Recent developments in air- and structure-borne sound and vibration*. Auburn University, U.S.A.

Boisson, C., Guyader, J.-L. & Lesueur, C. 1985 Etude numérique de la transmission d'énergie vibratoire entre structures assemblées: cas d'assemblages en L, T et al. *Acustica* **58**, 223–233.

Craik, R. J. M., Steel, J. A. & Evans, D. I. 1991 Statistical energy analysis of structure-borne sound transmission at low frequencies. *J. Sound Vib.* **144**, 95–107.

Davies, H. G. 1981 Ensemble averages of power flow in randomly excited coupled beams. *J. Sound Vib.* **77**, 311–321.

Dimitriadis, E. K. & Pierce, A. D. 1988 Analytical solution for the power exchange between strongly coupled plates under random excitation: A test of statistical analysis concepts. *J. Sound Vib.* **123**, 397–412.

Fahy, F. J. & Mohammed, A. D. 1992 A study of uncertainty in applications of SEA to coupled beam and plate systems, Part I: Computational experiments. *J. Sound Vib.* **158**, 45–67.

Fahy, F. J. 1970 Response of a cylinder to random sound in the contained fluid. *J. Sound Vib.* **13**, 171–194.

Fahy, F. J. & Yao De Yuan 1987 Power flow between non-conservatively coupled oscillators. *J. Sound Vib.* **114**, 1–11.

Gersch, W. 1968 Average power and power exchange in oscillators. *J. Acoust. Soc. Am.* **46**, 1180–1185.

Gibbs, B. M. & Gilford, C. L. S. 1976 The use of power flow methods for the assessment of sound transmission in building structures. *J. Sound Vib.* **49**, 267–286.

Hodges, C. H. & Woodhouse, J. 1986 Theories of noise and vibration transmission in complex structures. *Rep. Mod. Phys.* **49**, 107–170.

Keane, A. J. & Price, W. G. 1987 Statistical energy analysis of strongly coupled systems. *J. Sound Vib.* **117**, 363–386.

Keane, A. J. & Price, W. G. 1989*a* On the vibrations of mono-coupled periodic and near-periodic structures. *J. Sound Vib.* **128**, 423–450.

Keane, A. J. & Price, W. G. 1989*b* Statistical energy analysis of periodic structures. *Proc. R. Soc. Lond.* A **423**, 331–360.

Keane, A. J. & Price, W. G. 1991 A note on the power flowing between two conservatively coupled multi-modal systems. *J. Sound Vib.* **144**, 185–196.

Keane, A. J. & Price, W. G. 1992 Energy flows between arbitrary configurations of conservatively-coupled multi-modal elastic subsystems. *Proc. R. Soc. Lond.* A **436**, 537–568.

Kompella, M. S. & Bernhard, B. J. 1993 Measurement of the statistical variation of structural-acoustic characteristics of automotive vehicles. In *Proc. SAE Noise and Vibration Conf.* Warrendale, U.S.A.: Society of Automotive Engineers.

Lalor, N. 1991 Energy flow analysis for the diagnosis and cure of NVH problems. Paper no. 911323. In *Proc. 24th Int. Symp. on Automotive Technology and Automation.* Florence, Italy: ISATA.

Langley, R. S. 1989 A general derivation of the statistical energy analysis equations for coupled dynamic systems. *J. Sound Vib.* **135**, 499–508.

Lyon, R. H. & Maidanik, G. 1962 Power flow between linearly coupled oscillators. *J. Acoust. Soc. Am.* **34**, 623–629.

Lyon, R. H. 1975 *Statistical energy analysis of dynamical systems, theory and applications.* Massachusetts, U.S.A. MIT Press.

Mace, B. R. 1992 Power flow between two continuous one-dimensional subsystems; a wave solution. *J. Sound Vib.* **154**, 289–319.

Maidanik, G. 1976 Variations in the boundary conditions of coupled dynamic systems. *J. Sound Vib.* **46**, 585–589.

Newland, D. E. 1969 Power flow between a class of coupled oscillators. *J. Acoust. Soc. Am.* **43**, 553–559.

Norton, M. P. & Fahy, F. J. 1988 Experiments on the correlation of dynamic stress and strain with pipe wall vibrations for statistical energy analysis applications. *Noise Control Engng. J.* **30**, 107–118.

Plunt, J. 1980 Methods for predicting noise levels in ships. Ph.D. thesis, Chalmers University of Technology, Gothenburg, Sweden.

Roozen, N. B. 1992 Quiet by design: numerical acousto-elastic analysis of aircraft structures. Ph.D. thesis, Technical University of Eindhoven, The Netherlands.

Scharton, T. D. & Lyon, R. H. 1968 Power flow and energy sharing in random vibration. *J. Acoust. Soc. Am.* **43**, 1332–1343.

Scharton, T. D. 1979 Frequency-averaged power flow into a one-dimensional acoustic system. *J. Acoust. Soc. Am.* **50**, 373–381.

Smith, P. W. 1979 Statistical models of coupled dynamical systems and the transition from weak to strong coupling *J. Acoust. Soc. Am.* **65**, 695–698.

Stearn, S. M. 1970 Spatial variation of stress, strain and acceleration in structures subject to broad frequency band excitation. *J. Sound Vib.* **12**, 85–97.

Woodhouse, J. 1981 An approach to the theoretical background of statistical energy analysis. *J. Acoust. Soc. Am.* **69**, 1695–1709.

Zeman, J. L. & Bogdanoff, J. L. 1969 A comment on complex structural response to random vibrations. *AIAA J.* **7**, 1225–1231.

Statistical energy analysis as a tool for quantifying sound and vibration transmission paths

By M. Heckl and M. Lewit

Institut für Technische Akustik, Technische Universität, Berlin, Germany

When a complex structure is excited in several different ways by different sources, the SEA energy balance equations result in a set of linear equations that can be used to calculate loss factors, coupling loss factors or net energy flows and incoming powers. If certain symmetry relations are used, and/or if some prior knowledge about the system is available, the set of linear equations is overdetermined and can be solved by a least square technique.

A good indicator for the direction of the energy flow is the SEA temperature of the subsystems.

Experiments and computer simulations performed on three plate arrangements gave in general good results when the coupling was weak and there were more than three modes in the frequency band of interest. Not so good results were obtained when a small energy flow has to be measured as the difference of large quantities.

1. Introduction

Statistical energy analysis (SEA) was developed more than thirty years ago by Lyon, Smith, Maidanik, and others as a tool for predicting the mean square velocities of thin space-craft or aircraft structures when they are excited by sources (jet noise, turbulent boundary layer, etc.) that are random in nature and therefore contain wide frequency bands. A comprehensive description of the basic ideas, and some applications of SEA, is given in a book by Lyon (1975).

As its forerunner, the heat conduction model for the vibration distribution in buildings (Westphal 1957), SEA consists of a system of linear equations that describe the energy flow between substructures of a complex system. Because those equations are manifestations of the law of conservation of energy they are very robust. Quite often they give good results even if the usual requirements for their validity (large number of modes, sufficient modal overlap, etc.) may be violated.

As SEA has been successful in the prediction of average vibration amplitudes and sound pressures in space vehicles, airplanes, ships, buildings, large machines, etc., it certainly is worthwhile to try to use it as a tool for solving the 'inverse problem'; i.e. to investigate the energy flow and the coupling properties in existing structures. This would be of considerable help for optimizing noise control. It would allow us to find those paths that are responsible for the sound transmission in complex arrangements. It also would be of considerable help in the appropriate design of additional damping and isolation.

2. Possible applications of the 'inverse' SEA

In buildings, ships, vehicles, machines, etc., the sound is very often caused by several sources and is transmitted along different paths. Because the sound powers

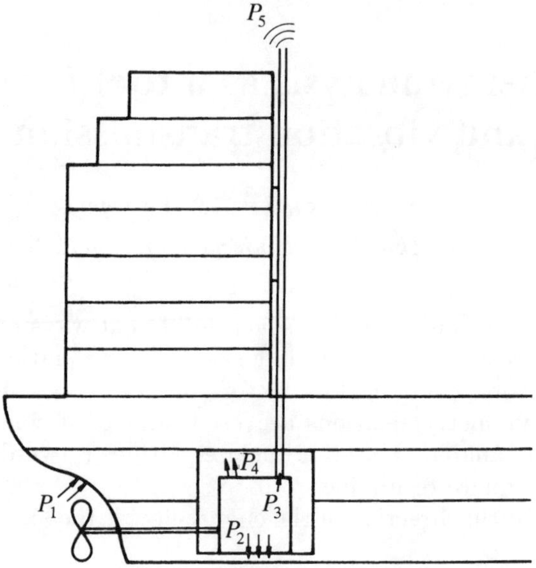

Figure 1. Major sources of acoustic power in a ship.

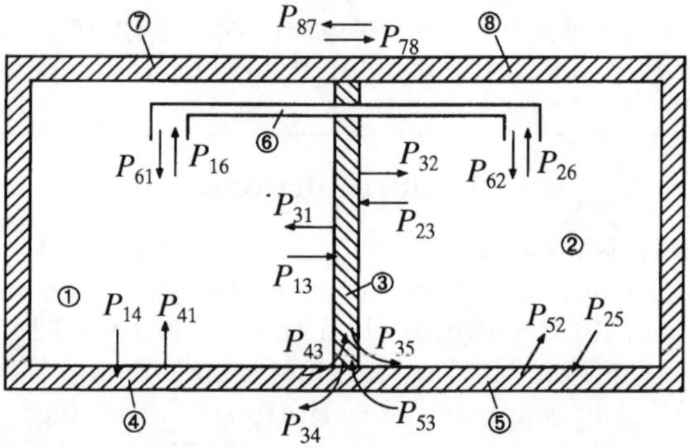

Figure 2. Acoustic energy flow between two adjacent rooms in a building. 1, Air-filled space (source room); 2, air-filled space (receiver room); 3, partition; 4, 5, 7, 8, flanking walls; 6, air duct. The arrows indicate the energy flow (not all possibilities are shown).

that are transmitted into a structure, appear explicitly in the SEA equations, it should be possible to determine them if all the other quantities in the equations are known. Thus if one wants, for example, to measure (see figure 1) the powers that are generated by a ship propeller (P_1), the main engine vibrations (P_2), the exhaust pipe vibrations (P_3), the air-borne sound from the engine (P_4), the radiated exhaust noise (P_5), one should be able to do so by determining the mechanical energies

$$E_\nu = m_\nu v_\nu^2 \quad \text{or} \quad E_\nu = (V_\nu/\rho c)\, p_\nu^2 \tag{1}$$

in certain parts of the ship and inserting them into the SEA equations. In (1) E_ν is the mechanical energy in the νth subsystem of mass m_ν or volume V_ν. The parameters ρ, c are the density and speed of sound in the air- or liquid filled volume V_ν. The value v_ν^2 is the mean square velocity of the νth subsystem and p_ν^2 the mean square sound pressure. Obviously the powers can only be measured if all the loss factors and coupling loss factors that appear in the SEA equations are known (which is hardly ever the case).

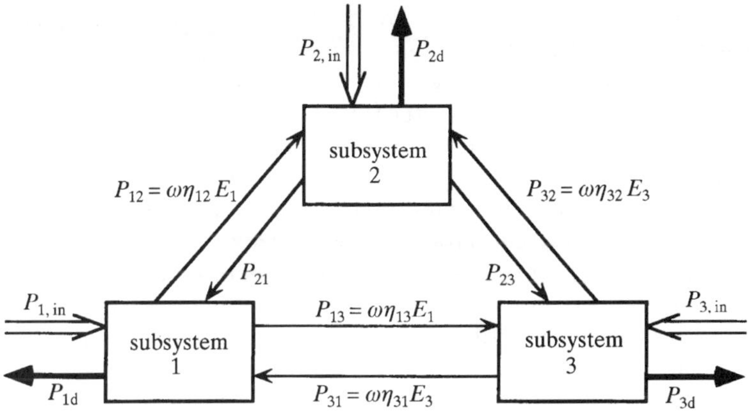

Figure 3. Idealization of a structure consisting of three subsystems (see equation (2)).

Figure 2 shows an example from building acoustics. The aim in this case may be to quantify the powers $P_{k\nu}$ that flow from one system to another to find out where sound is mainly transmitted and where any further isolation is most effective.

In this paper we are interested in the transmission paths; i.e. the energy flow quantities $P_{k\nu}$ which in standard SEA are usually expressed in terms of the coupling loss factors $\eta_{k\nu}$.

To measure these quantities in a three subsystem arrangement (see figure 3) we can make the following sequence of experiments.

In the first experiment subsystem 1 is excited by a stationary random power $P_1^{(1)}$ and the average energies $E_\nu^{(1)}$ are measured in each subsystem ($\nu = 1,2,3$). In the second experiment subsystem 2 is excited by $P_2^{(2)}$ and $E_\nu^{(2)}$ is determined. In a third experiment one proceeds in a similar way. This allows us to set up the following equation:

$$\left.\begin{aligned} &\eta_{\nu d}E_\nu^{(\mu)} + \sum_{k=1,\,\neq\nu}^{3} (\eta_{\nu k}E_\nu^{(\mu)} - \eta_{k\nu}E_k^{(\mu)}) = \frac{1}{\omega}P_{\nu,\text{in}}^{(\mu)}\,\delta_{\mu\nu} \\ \text{for}\quad &\nu = 1,2,3, \quad \mu = 1,2,3, \quad \delta_{\nu\nu} = 1; \quad \delta_{\nu\mu} = 0 \quad \text{for} \quad \nu \neq \mu. \end{aligned}\right\} \tag{2}$$

Here the superscript indicates the number of the experiment and the subscript gives the number of the subsystem. ω is the angular frequency, $\eta_{1d}, \eta_{2d}, \eta_{3d}$ are the loss factors that characterize powers

$$P_{\nu d} = \omega\eta_{\nu d}E_\nu \tag{3}$$

lost in each system. $\eta_{\nu k}$ are the coupling loss factors; they determine the energy flow (power) from the νth subsystem to the k-subsystem; i.e.

$$P_{\nu k} = \omega\eta_{\nu k}E_\nu. \tag{4}$$

As (2) consists of nine linear equations, one might argue that it allows to determine nine unknown quantities; i.e. the six coupling loss factors $\eta_{\nu k}$, and the three internal loss factors $\eta_{\nu d}$ or the $\eta_{\nu k}$ values and the input powers $P_\mu^{(\mu)}$, etc.

If the number of subsystems is not three, the same procedure which always give n^2 linear equations can be applied (n = number of subsystems).

3. Short literature survey

The idea to use the SEA-equations to determine coupling loss factors is not new.

Apart from some general remarks in Lyon's book, Bies & Hamid (1980) seem to have been the first to use SEA in an inverse way. They realized that the accuracy of the method is not very high, and therefore used an overdetermined system which they solved by an error minimization technique.

Woodhouse (1981) also discussed the accuracy problem. He suggested modifying the measured data in such a way that the final results do not violate the basic conservation law (i.e. no negative coupling loss factors are allowed). Obviously the modifications of the measured data were kept to a minimum. The numerical procedures for this method are described in more detail by Hodges *et al.* (1987). Clarkson & Ranky (1984) also found that *in situ* measurements of internal and coupling loss factors are of limited accuracy. They suggested an iteration scheme to improve the results.

Lalor *et al.* (1989, 1990) combined the equations in (2) to get

$$\left.\begin{aligned} \eta_{21}\left(\frac{E_2^{(2)}}{E_1^{(2)}}-\frac{E_2^{(1)}}{E_1^{(1)}}\right)+\eta_{31}\left(\frac{E_3^{(2)}}{E_1^{(2)}}-\frac{E_3^{(1)}}{E_1^{(1)}}\right)&=\frac{P_1^{(1)}}{\omega E_1^{(1)}},\\ \eta_{21}\left(\frac{E_2^{(3)}}{E_1^{(3)}}-\frac{E_2^{(1)}}{E_1^{(1)}}\right)+\eta_{31}\left(\frac{E_3^{(3)}}{E_1^{(3)}}-\frac{E_3^{(1)}}{E_1^{(1)}}\right)&=\frac{P_1^{(1)}}{\omega E_1^{(1)}}. \end{aligned}\right\} \tag{5a}$$

In a similar way they found

$$\left.\begin{aligned} \eta_{12}\left(\frac{E_1^{(1)}}{E_2^{(1)}}-\frac{E_1^{(2)}}{E_2^{(2)}}\right)+\eta_{32}\left(\frac{E_3^{(1)}}{E_2^{(1)}}-\frac{E_3^{(2)}}{E_2^{(2)}}\right)&=\frac{P_2^{(2)}}{\omega E_2^{(2)}},\\ \eta_{12}\left(\frac{E_1^{(3)}}{E_2^{(3)}}-\frac{E_1^{(2)}}{E_2^{(2)}}\right)+\eta_{32}\left(\frac{E_3^{(3)}}{E_2^{(3)}}-\frac{E_3^{(2)}}{E_2^{(2)}}\right)&=\frac{P_2^{(2)}}{\omega E_2^{(2)}} \end{aligned}\right\} \tag{5b}$$

and

$$\left.\begin{aligned} \eta_{13}\left(\frac{E_1^{(1)}}{E_3^{(1)}}-\frac{E_1^{(3)}}{E_3^{(3)}}\right)+\eta_{23}\left(\frac{E_2^{(1)}}{E_3^{(1)}}-\frac{E_2^{(3)}}{E_3^{(3)}}\right)&=\frac{P_3^{(3)}}{\omega E_3^{(3)}},\\ \eta_{13}\left(\frac{E_1^{(2)}}{E_3^{(2)}}-\frac{E_1^{(3)}}{E_3^{(3)}}\right)+\eta_{23}\left(\frac{E_2^{(2)}}{E_3^{(2)}}-\frac{E_2^{(3)}}{E_3^{(3)}}\right)&=\frac{P_3^{(3)}}{\omega E_3^{(3)}}. \end{aligned}\right\} \tag{5c}$$

By adding the equations for $\mu = 1, 2, 3$ in (2) one finds for the internal loss factors

$$\left.\begin{aligned} \eta_{1d}E_1^{(1)}+\eta_{2d}E_2^{(1)}+\eta_{3d}E_3^{(1)}&=P_1^{(1)}/\omega,\\ \eta_{1d}E_1^{(2)}+\eta_{2d}E_2^{(2)}+\eta_{3d}E_3^{(2)}&=P_2^{(2)}/\omega,\\ \eta_{1d}E_1^{(3)}+\eta_{2d}E_2^{(3)}+\eta_{3d}E_3^{(3)}&=P_3^{(3)}/\omega. \end{aligned}\right\} \tag{5d}$$

This way the coupling loss factors are separated from the internal loss factors and the equations that have to be solved are much simpler. Ming *et al.* (1990) applied this method to a car and found very reliable results.

Obviously the procedure underlying (5*a*–*d*) can also be applied to subsystems that are excited in n different ways. In this case the basic $n \times n$ equations can be rearranged so that there are n sets of $(n-1)$ equations containing the $n-1$ coupling loss factors. In addition there are n equations containing only the internal loss factors. Measuring all the n^2+n coefficients that are needed in these equations is a very formidable task, because for each frequency band averages have to be taken over the surface of each subsystem and also over several excitation points on each subsystem.

The rearrangement of (2) proposed by Lalor *et al.* simplifies the equations somewhat, but it also shows that for a strongly coupled system the method becomes very sensitive to the slightest error. When a system consists of strongly coupled subsystems it vibrates more or less the same way whichever subsystems are excited. Thus the ratio of energies in different experiments is practically equal – e.g. $E_1^{(1)}/E_2^{(1)} \approx E_1^{(2)}/E_2^{(2)}$ – and therefore the elements of the matrices of (5*a*–*c*) become almost zero.

4. The thermodynamic analogy as an indicator for the direction of energy flow

In the early publications on SEA it was mentioned already that there exists a thermodynamic analogy where the energy per mode corresponds to temperature and the coupling coefficient (which is not identical with the coupling loss factor) corresponds to the heat conduction coefficient. With this analogy in mind one can easily find the direction of energy flow in a construction composed of many multimodal substructures, because energy always flows from the higher temperature to the lower one. It is also obvious that subsystems are strongly coupled if their temperatures are equal or almost equal.

To apply this general idea it is necessary to find the SEA 'temperature' of the νth substructure. It is given by

$$T_\nu = E_\nu/\Delta N_\nu. \tag{6}$$

Here the energy E_ν is given by (1) and ΔN_ν is the number of modes of the ν-subsystem within the frequency band of interest.

The measurement of the energies E_ν can be done by standard techniques; the number of modes has to be estimated somehow. One method to obtain ΔN_ν would be to measure the number of resonances for each subsystem in the frequency bands of interest. When there are not too many resonances (average distance between two resonances larger than three times the bandwidth of a resonance) this may give a reasonably good value for ΔN_ν, even though it might not always be easy to decide whether a small peak must be considered as a resonance or not.

Another method is to use the asymptotic relation (Cremer *et al.* 1973)

$$\Delta N_\nu = 4m_\nu \operatorname{Re}\{1/Z_\nu\}\,\Delta f. \tag{6a}$$

Here Δf is the frequency range (in Hertz) of interest and Z_ν is the input impedance of the system, provided it is in the average not much influenced by boundary effects. $\operatorname{Re}\{1/Z_\nu\}$ may be taken from the literature 'for the corresponding infinite system', or it may be measured.

Figure 4 shows in its upper part on a logarithmic scale the SEA temperature that was measured on a three-plate rearrangement. Plate 2 was excited successively at five points, and 18 to 30 response locations per plate were used. All the measured input powers and energies were summed up. For the number of modes the asymptotic value for plates was taken.

In its lower part figure 4 gives the net energy flow $W_{\nu k}$ (see (8)) from one subsystem to another one using the methods described later. The power coming from outside was always $P_2^{(2)} = 1$ W, corresponding to 120 dB. Since W_{12}, W_{23}, W_{13} are net energy flows they may be positive or negative. To indicate their direction in the figure and still retain the logarithmic scale, positive and negative energy flows are plotted in

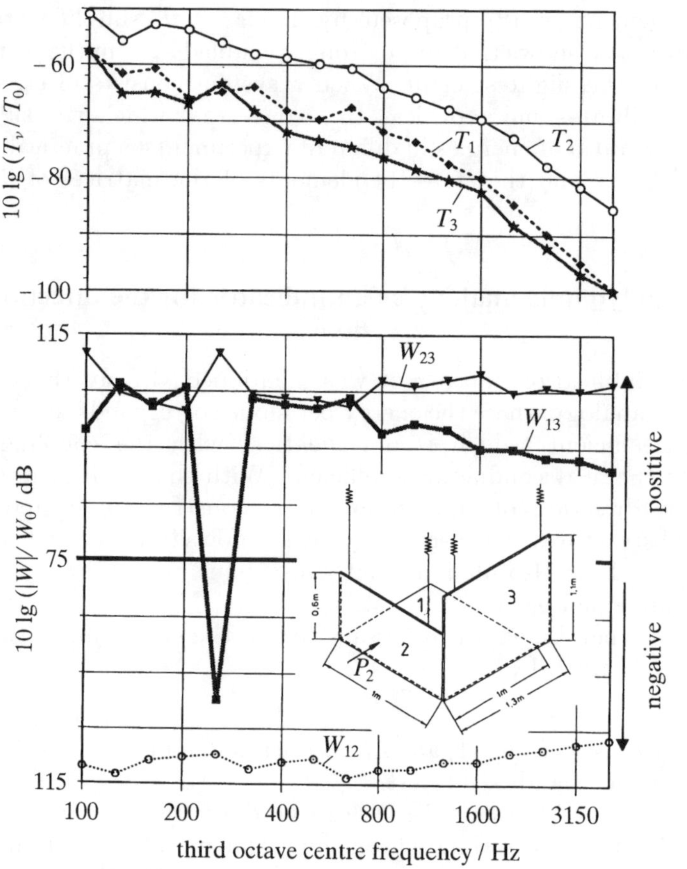

Figure 4. Temperature and net energy flow in a three plate system.

different directions. The range below 75 dB is missing (it is not known accurately anyway). The reference value always is $W_0 = 10^{-12}$ W.

It can clearly be seen that the direction of the SEA-measured energy flow agrees with the sign of the temperature difference although the plates did not have a high modal density. In the third octave centred at 200 Hz there were approximately three modes, in the 2000 Hz third octave about 32.

Figures 5 and 6 show the results of computer simulations. In this case the modal expansion of three coupled simply supported plates was used to calculate the mean square velocities of the plates. The parameter of the plates were chosen in such a way that the asymptotic number of modes was the same as in the example shown in figure 4 (0.058 to 0.064 modes Hz^{-1}). The velocities for three different types of excitation which were found this way were then introduced into (2) to obtain the net energy flow. In the example shown in figure 5, plate 1 was excited, but because there was a soft spring between 1 and 2 and a stiff spring between 1 and 3 the SEA temperature T_3 is higher than T_2 and consequently W_{23} is negative; i.e. there is a net energy transport from 3 to 2. With respect to the energy flow from 1 to 2 the temperature curves clearly indicate a flow from 1 to 2, which is quite plausible. The curve for W_{12}, however, gives partly positive and partly negative values. This is probably due to the fact that W_{12} is rather small and therefore its measurement is not very reliable. In computer simulated experiments, the energy flows can also be calculated directly to

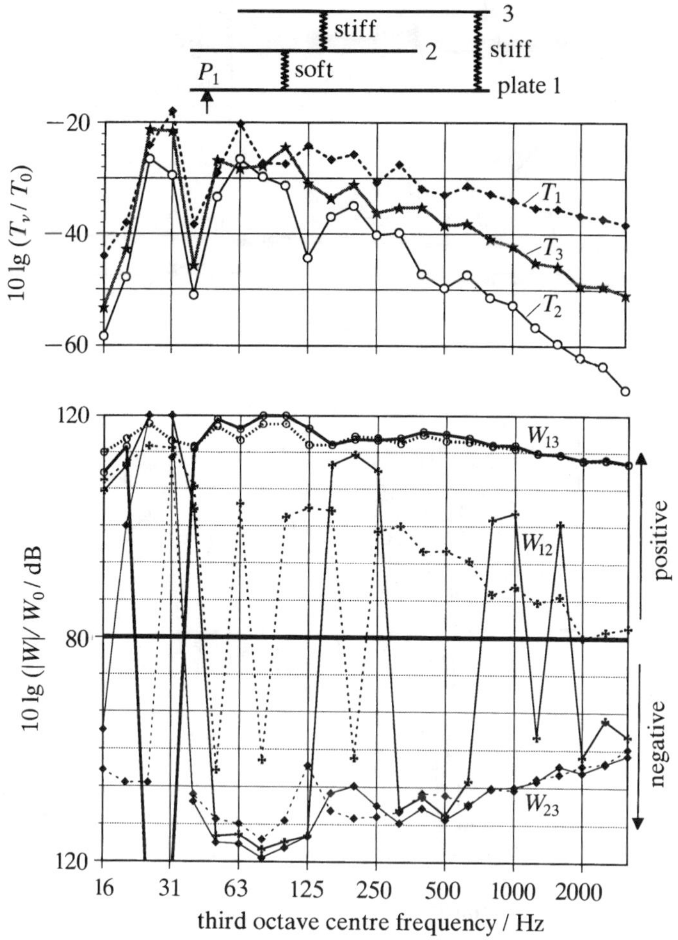

Figure 5. Temperature and net energy flow in a three plate system (computer simulation). Broken lines are the true values based on modal expansion. Continuous lines are the SEA calculations based on simulated experiment.

verify the SEA-measured values. They are included in figures 5 and 6, and it can be seen that the results are very good for the larger energy flows above 125 Hz.

In the example shown in figure 6 the centre plate 2 was excited and all coupling springs were very stiff. The SEA temperature curves indicate an energy flow from 2 to 3 and 2 to 1. This agrees with the curves for W_{23} and $W_{12} = -W_{21}$. Above 125 Hz T_1 is in general greater than T_3 and therefore W_{13} is mainly positive.

For frequencies below 125 Hz the curves for the net energy flow are rather erratic because in this frequency range the system is well coupled and there are only two or less modes in a third octave band.

In conclusion, this part of the paper shows that the measurement of the SEA temperature (based on the asymptotic number of modes) yields a reasonably good indication of the net energy flow direction. In addition, the SEA temperature distribution helps to find those subsystems that are strongly coupled because their SEA temperatures are more or less equal.

Measurements of the SEA temperature are especially useful when subsystems of completely different types are connected. Examples are the sound pressure in a small

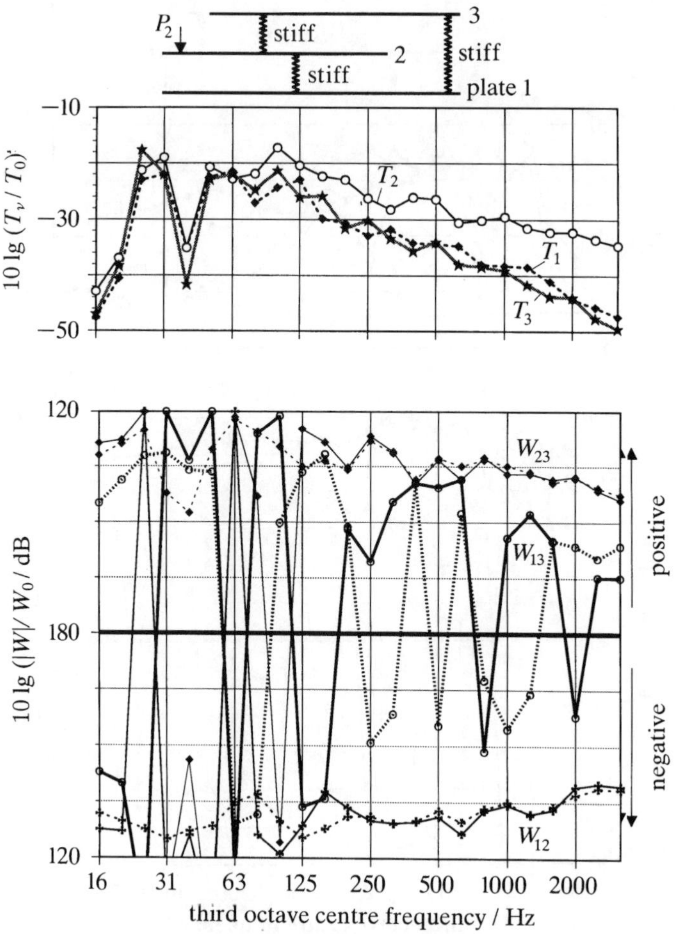

Figure 6. See figure 5.

space (e.g. the oil in a pipe) which is coupled to the vibrations of the surrounding walls, or the in-plane waves in a plate that are coupled to the bending waves of another (or the same) structure. In such cases the measured quantities such as root mean squared velocities, accelerations, or pressures cannot be compared, but the SEA temperatures can.

5. Measurements and computer simulations

(a) *Internal loss factors*

For the measurement of internal loss factors the SEA equations are added in such a way that all terms containing the coupling loss factors are eliminated. For a three subsystem arrangement (5d) is obtained this way. In the general case of n subsystems which are excited successively by n outside forces, the resulting set of equations is

$$\sum_{\nu=1}^{n} \eta_{\nu d} E_\nu^{(\mu)} = \frac{1}{\omega} P_{\mu,\,\text{in}}^{(\mu)} \delta_{\nu\mu}. \tag{7}$$

Here the number of different excitation situations has to range from $\mu = 1$ to $\mu = n$.

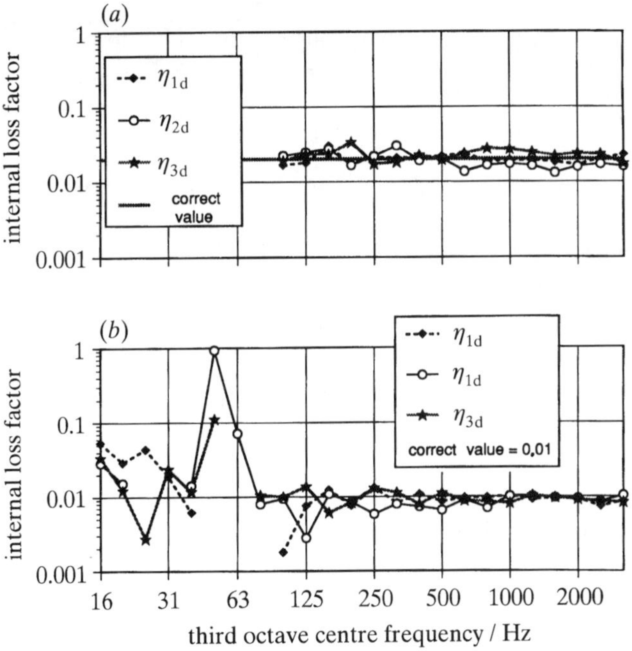

Figure 7. Internal loss factors calculated by using equation (7). (*a*) Real experiments using three plates. (*b*) Computer simulation. (In both cases there were less than two modes per frequency band below 125 Hz.)

This method was applied successfully by Ming *et al.* (1990). Other results are shown in figure 7. They are reported by Lewit & Lehmköster (1992).

In the 'real' experiment three wood fibre boards (see also figure 4), which had different sizes and thicknesses but the same loss factors, were rigidly connected at the edges and $\eta_{1d}, \eta_{2d}, \eta_{3d}$ was measured by using (7). The 'correct value', which is also given in figure 7, is based on the reverberation times at different frequencies of a free-free bar that consisted of the same material. The computer simulation in figure 7 is based on a three plate configuration. In this case the mean square velocities were calculated using a modal expansion. The results were then inserted into (7).

If one keeps in mind that the results are based on measured average levels that have an accuracy of 1–2 dB, the agreement is surprisingly good.

(*b*) *Net energy flow*

In (2) the energy flow between two subsystems is expressed in terms of the loss factor. But in the literature the basic SEA relations are also written as

$$\left.\begin{aligned} \omega\eta_{\nu d}E_\nu + \sum_{k\neq\nu} W_{\nu k} &= P_{\nu,\,\mathrm{in}} \\ \text{or}\qquad \eta_{\nu d}E_\nu + \sum_{k\neq\nu} \alpha_{\nu k}(T_\nu - T_k) &= \frac{1}{\omega}P_{\nu,\,\mathrm{in}}. \end{aligned}\right\} \tag{8}$$

Here $W_{\nu k}$ is the net energy flow from ν to k. The values T_ν and T_k are the SEA temperatures given by (6). A very important aspect for the application of (8) is the symmetry relation for the 'conductances' $\alpha_{\nu k}$; i.e.

$$\alpha_{\nu k} = \alpha_{k\nu}. \tag{9}$$

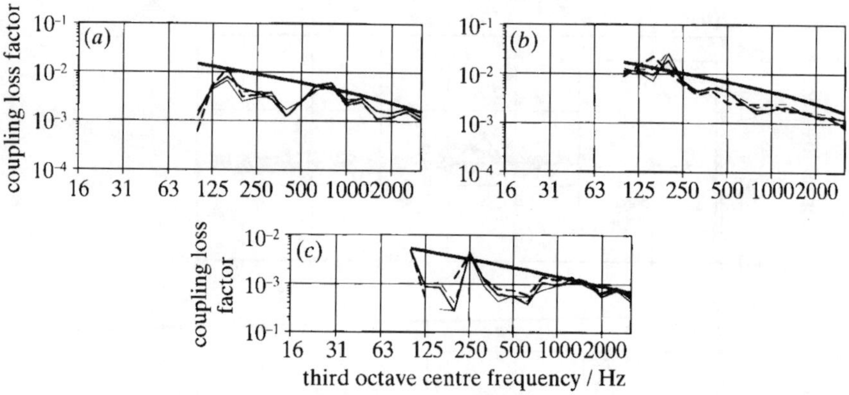

Figure 8. Comparison of measured and theoretical (thick) coupling loss factors. (a) η_{12}, (b) η_{13}, (c) η_{23}.

In the basic SEA theory this relation is proved for coupled oscillators. If (8) is applied to n different excitation situations, the result is

$$\eta_{\nu d} E_{\nu}^{(\mu)} + \sum_{k \neq \nu} \alpha_{\nu k}(T_{\nu}^{(\mu)} - T_{k}^{(\mu)}) = \frac{1}{\omega} P_{\mu,\,\mathrm{in}}^{(\mu)} \delta_{\nu\mu}. \tag{10}$$

Here $P_{\mu,\,\mathrm{in}}^{(\mu)}$ is again the power that is transmitted from outside into the μth subsystem (and only into this one) in the μth experiment.

Equation (10) is a system of $n \times n$ equations. By adding those relations that have the same value of μ, we again find (7). Because we used this relation already for finding the internal loss factors, we can exclude n equations from (8). For the sake of simplicity we choose to exclude those relations that have the incoming powers on their right-hand side. This way we are left with $n(n-1)$ relations that do not contain the incoming powers (which are sometimes hard to measure); i.e.

$$\sum_{k \neq \nu} \alpha_{\nu k}(T_{\nu}^{(\mu)} - T_{k}^{(\mu)}) = -\eta_{\nu d} E_{\nu}^{(\mu)}. \tag{11}$$

Here ν and μ range from 1 to n but $\nu = \mu$ is excluded. Because of the symmetry relation $\alpha_{\nu k} = \alpha_{k\nu}$, equation (11) is an overdetermined system of $n(n-1)$ equations for $\frac{1}{2}n(n-1)$ unknowns. Thus we would be allowed to exclude half of the equations in (11). But as there is no simple rule for deciding which should be retained, it is better to write (11) as

$$\sum_{k \neq \nu} \alpha_{\nu k} D_{\nu k}^{(\mu)} = -\eta_{\nu d} \Delta N_{\nu} \tag{12}$$

and to solve this overdetermined system by a least square technique. In (11) the relative temperature difference

$$D_{\nu k}^{(\mu)} = (T_{\nu}^{(\mu)} - T_{k}^{(\mu)})/T_{\nu}^{(\mu)} \tag{13}$$

was introduced, where ΔN_{ν} is again the number of modes in the νth subsystem. The relative temperature difference constitutes a normalization which helps to avoid $P_{\mu,\,\mathrm{in}}$ having a strong influence on the final results. Figure 8 shows results that were obtained by this and other methods from the setup introduced in figure 4. The quantity plotted in figure 8 is $\eta_{\nu k} = \alpha_{\nu k}/\Delta N_{\nu}$.

Apart from experiments with the real three plate arrangement, several computer simulations were made using the model that is briefly described in connection with figures 5 and 6. By using the modal expansion the 'true' net energy flow could be calculated and compared with the energy flow that is found from (12). The result is shown in the upper part of figure 10. It can be seen that there is good agreement between the 'true value' and the SEA calculation based on simulated experiments.

(*c*) *Coupling loss factors*

If the SEA equations are applied to several excitation situations (e.g. (2)) it is useful to distinguish between the following two cases:

1. No assumption is made with respect to a relation between the coupling loss factors $\eta_{\nu k}$ and $\eta_{k\nu}$. In this case one has to solve the full set of linear equations

$$\eta_{\nu d} E_\nu^{(\mu)} + \sum_{k \neq \nu} (\eta_{\nu k} E_\nu^{(\mu)} - \eta_{k\nu} E_k^{(\mu)}) = \frac{1}{\omega} P_{\mu,\,\mathrm{in}}^{(\mu)} \delta_{\nu\mu} \tag{14}$$

with $1 \leqslant \nu \leqslant n$, $1 \leqslant \mu \leqslant n$. It can be seen that there are n^2-equations and the same number of unknowns. Thus the system can be solved either directly or after making the Lalor-rearrangement that was used in (5*a*–*d*).

2. If, however, the reciprocity relation

$$\eta_{\nu k} \Delta N_\nu = \eta_{k\nu} \Delta N_k, \tag{15}$$

which is rather fundamental in SEA, is applied, the number of unknowns is reduced to $n + \frac{1}{2}n(n-1)$, because, apart from the n values for $\eta_{\nu d}$, we have to find only $\frac{1}{2}n(n-1)$ coupling loss factors $\eta_{\nu k}$.

There are several ways to treat this overdetermined system.

(i) Because the absolute measurement of $P_{\mu,\,\mathrm{in}}$ is generally of limited accuracy, all equations containing this quantity are excluded (see Lewit & Petit 1991). This way an overdetermined, homogeneous system of equations is obtained. As coefficients it contains the energy ratios E_k/E_ν. With a least square technique

$$\eta_{\nu d}/\eta_{\mathrm{ref}} \quad \text{and} \quad \eta_{\nu k}/\eta_{\mathrm{ref}}$$

can be found. The quantity η_{ref}, for which one may take η_{1d} or any other loss factor, is the only loss factor that cannot be calculated by this method which is based on a homogeneous set of equations. Usually this is not a serious drawback because relative loss factors contain most of the desired information. If this should not be the case, at least one non-zero loss factor has to be found by an independent method.

(ii) The assumption (15) is equivalent to setting

$$\eta_{\nu k} = \alpha_{\nu k}/\Delta N_\nu. \tag{16}$$

This way – except for a factor – equation (12) is obtained and the methods described there can be applied.

(iii) In principle one could exclude $\frac{1}{2}n(n-1)$ equations from (14). But it is hard to decide what can be excluded, without getting an ill-conditioned matrix or other accuracy problems; therefore this method is not discussed further here.

Figure 8 shows results that were obtained by using the different methods. The data are so close together that it did not seem necessary to indicate which curve belongs to which method. In addition, figure 8 also gives a theoretical solid line (W. Wöhle, personal communication). It is based on the theoretical transmission coefficient of two infinite plates that are connected at right angles.

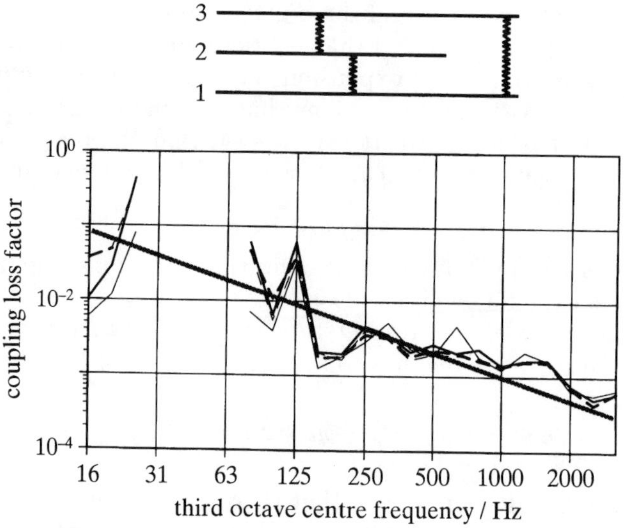

Figure 9. Comparison of coupling loss factors derived from computer simulations and theory (thick line). Internal loss factor 0.01.

In figure 9 the results of computer simulations are plotted. In this case three plates already introduced in figure 6 were coupled by rigid connectors; therefore a theoretical coupling loss factor can be derived from Cremer's (1953) formula for sound bridges between two plates. The result is

$$\eta_{\nu k} = \frac{1}{2|A|^2 \omega m_\nu} \mathrm{Re}\left\{\frac{1}{Z_k}\right\}, \quad \text{with} \quad A = \frac{1}{Z_\nu} + \frac{1}{Z_k}.$$

Here Z_ν, Z_k are the impedances of subsystems ν and k if they were infinite.

When the experimental data and the theoretical curve in figures 8 and 9 are compared, one is confronted with the question whether the discrepancies are due to shortcomings of the inverse SEA method or due to the inapplicability of the theoretical data to the experimental situation.

It is our opinion that – even above 160 Hz when there are at least three modes per third octave and the systems are weakly coupled – the theoretical data do not adequately describe the experimental situation and therefore the discrepancies do not indicate shortcomings of the method. The reasons for this opinion are as follows.

(*a*) The general trend of the theoretical and experimental curves is the same.

(*b*) Measurements (which are not repeated here) showed that within 1–2 dB the measured coupling loss factors obeyed the reciprocity relation $\eta_{12} \cdot \eta_{23} \cdot \eta_{31} = \eta_{21} \cdot \eta_{32} \cdot \eta_{13}$ (obviously this test was only made when (15) was not used).

(*c*) The four different methods (see figure 9) gave practically the same results.

(*d*) In the computer simulation the modal expansion allows to calculate the 'true' net energy flow between two finite systems. It can also be used to simulate the inverse SEA method which is based on squared quantities. In figure 10*a* comparison is made for the arrangement from figure 6 but with very low internal loss factors (0.0001). It shows that the net energy flow W_{12} is determined rather exactly although the measured coupling loss factors do not agree with the theoretical values. Thus agreement with theoretical values that are based on infinite substructure transmission coefficients cannot be required.

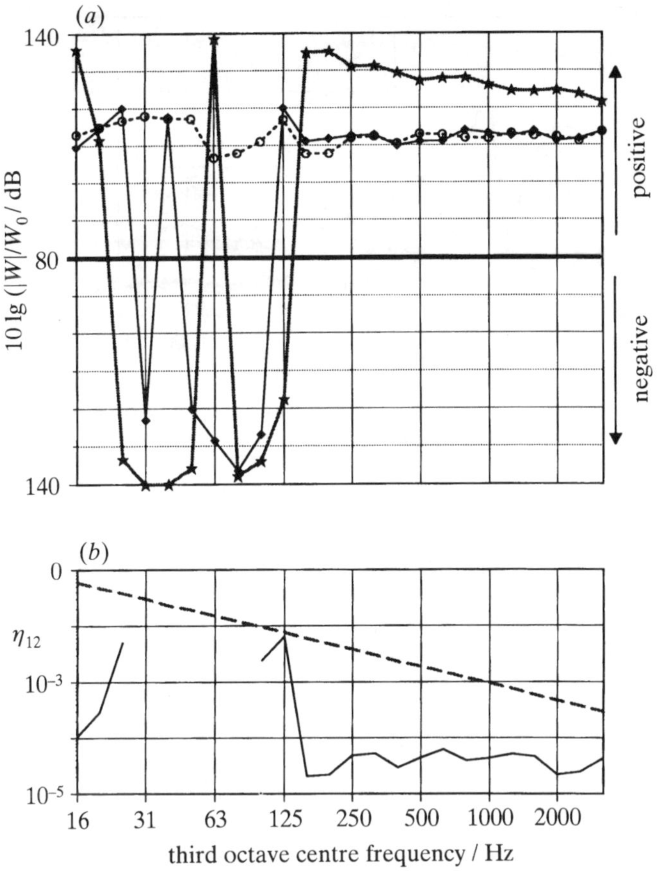

Figure 10. (*a*) Net energy flow, W_{12}, of a three plate arrangement (computer simulation). ○, True value based on modal expansion; ◆, SEA calculation based on simulated experiment; ★, SEA calculation based on 'theoretical' coupling loss factor. (*b*) Coupling loss factor, η_{12} of a three plate arrangement. ——, Result of simulated SEA calculation; ---, 'theoretical' (infinite) value. Internal loss factor 0.0001.

(*e*) For an investigation of the applicability of 'theoretical' values it is very revealing to study the influence of internal damping on the coupling loss factors. There are two limiting cases: (i) when two simple degree of freedom resonators are coupled, the coupling loss factors depend strongly on internal damping (see Lyon 1975 or Cremer *et al.* 1973); (ii) when two infinite subsystems (infinite number of modes) are coupled, internal damping has only a second order influence on the coupling loss factor. Practical situations are somewhere in between and indeed it could be shown by computer simulations, that the coupling loss factors depend on the internal damping of the receiving substructure whenever $\eta_{\nu k} \gg \eta_{\nu d}$; i.e. when there is strong coupling. The lower part of figure 10 gives an example of this type; it shows an 'experimental' coupling loss factor which is well below the one plotted in figure 9 because the internal loss factors are much smaller.

The upper part of figure 10 shows that the 'theoretical' coupling loss factor would even lead to a net energy flow W_{12} that is larger than the incoming power of 1 W → 120 dB.

The examples shown in most of the graphs indicate that SEA allows us to quantify

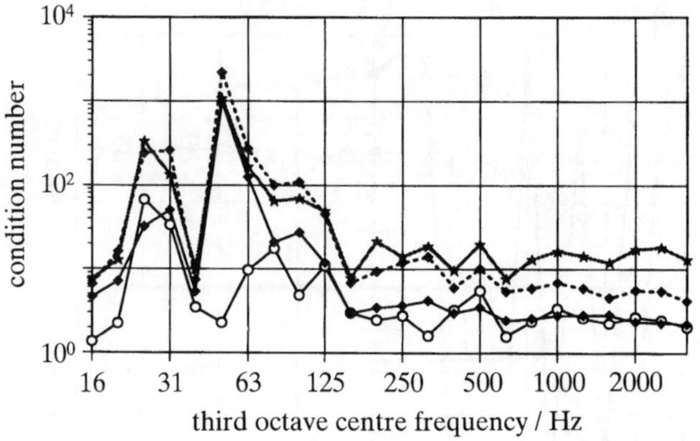

Figure 11. Condition number for different methods of measuring coupling loss factors. —◆—, Equation (14) without relations containing $P_\mu^{(\mu)}(n_E = 6, n_u = 6)$; —○—, equation (11) with $\eta_{\nu d}$ known ($n_E = 6, n_u = 3$); ★, equation (14) combined with equation (15) ($n_E = 6, n_u = 6$); —◆—, equation (5*a*–*d*) ($n_E = 9, n_u = 9$). n_E = number of equations, n_u = number of unknowns.

the transmission paths in a complex system. But obviously there are also drawbacks. Apart from the rather large experimental effort, a fundamental one is that small net energy flows may be buried under larger ones. Figure 5 (lower part) contains an example of this type. Here the curve for W_{12} is rather erratic (and in disagreement with the true energy flow calculation) although the temperatures clearly show that energy must flow from plate 1 to plate 2. This result – and others not reported here – shows difficulties that arise when there is a small energy flow, derived from a small coupling loss factor, in the presence of a large energy flow from somewhere else into the same subsystem. But as one is usually interested mainly in the important transmission paths this is not too great a disadvantage.

(*d*) *Condition number*

Whenever one is dealing with an inverse problem, the question of ill-posedness comes up. For the case of linear equations as they appear in SEA, a problem is ill-posed when the condition number is high (see Stewart 1973). Thus it seems useful to accompany calculations of the type reported here with the determination of the condition number. Figure 11 shows several examples. It can be seen that at low frequencies, when there are only a few modes, the condition number became high; in the remaining frequency range it was between 1 and 10.

It does not seem possible to give a fixed upper limit for the condition number. But at least one can say that an error of $p(\%)$ in the input data leads at worst to an error of $C_n p(\%)$ in the final result (C_n is the condition number). Thus for $C_n = 10$ a 1 dB (= 25 %) error in the input data leads, in the worst case, to a result that deviates by 250 % (= 4 dB) from the correct value. Such errors may appear high but, because they are worst cases, it seems reasonable to assume that with some care the average accuracy of the final results can be brought to 2–3 dB.

6. Conclusions

The 'inverse' use of SEA constitutes a useful method for gaining information about the transmission paths in complex structures.

If only the direction of sound and vibration transmission is of interest, it is sufficient to determine the SEA temperatures according to (6). The net energy flow always is from the higher temperature to the lower one.

If two adjacent subsystems independent of source position have the same temperature, they are well coupled. Temperatures can be compared even if they are based on the mean square velocities of different wave types or on pressures in different fluids or gases.

When SEA is used to calculate the net energy flow, or the internal loss factor, or the coupling loss factors in a given complex system, then several experiments are required. They involve the excitation with different source configurations and the measurement of many mean square velocities (or pressures). The data obtained this way are used to establish a set of linear equations for the unknown quantities. There are several ways for solving this set of equations, especially when they form an overdetermined system. This is the case when the symmetry relations (9) or (15) are used or when certain quantities are known *a priori* from other independent measurements.

Practical applications and computer simulations on arrangements consisting of three plates generally gave good results when there were more than three modes in a frequency band of interest and when the coupling was weak. Poor results for the coupling quantities were obtained when (i) the corresponding subsystems were well coupled (i.e. small temperature difference), (ii) the total energy flow into a certain subsystem originates to a small part from one neighbouring system and to a large part from another one. In the first case the coupling loss factor is not the appropriate way to describe the situation (the temperature distribution would be better), the second case is not of great practical relevance.

Comparison of measured coupling loss factors with theoretical values that are based on calculation of the transmissibility of infinite substructures, should be made only when the subsystems are not well coupled, i.e. when they have sufficient damping. The conditions underlying such theoretical data are much more stringent than the conditions for the applicability of SEA.

The condition number of the matrix that has to be inverted always should be calculated, because it gives a good indication of the error sensitivity of the results.

References

Bies, D. A. & Hamid, S. 1990 In situ determination of loss and coupling loss factors by the power injection method. *J. Sound Vib.* **70**, 187–204.

Clarkson, B. L. & Ranky, M. F. 1984 On the measurement of the coupling loss factor of structural connections. *J. Sound Vib.* **94**, 249–261.

Cremer, L., Heckl, M. & Ungar, E. E. 1973 *Structure-borne sound*, pp. 474–490. New York: Springer-Verlag.

Cremer, L. 1954 Berechnung der Wirkung von Schallbrücken. *Acustica* **4**, 273–276.

Hodges, C. H., Nasch, P. & Woodhouse, J. 1987 Measurement of coupling loss factors by matrix fitting: Investigation of numerical procedures. *Appl. Acoustics* **22**, 47–69.

Lalor, N. 1989 The experimental determination of vibrational energy balance in complex structures, paper no. 108429. In *Proc. SIRA Conf. on Stress and Vibration.*

Lalor, N. 1990 Considerations for the measurement of internal and coupling loss factors on complex structure. *ISVR Tech. Report no. 182.*

Lewit, M. & Lehmköster, S. 1992 Probleme und Lösungswege in der Energieflußanalyse. *Fortschritte der Akustik DAGA*, pp. 1009–1012. Bad Honnef: DPG.

Lewit, M. & Petit, M.-F. 1991 Messung von Kopplungsverlustfaktoren und Eigenverlustfaktoren beliebiger Systeme in der Statistischen Energieanalyse. *Fortschritte der Akustik DAGA*, pp. 345–348. Bad Honnef: DPG.

Lyon, R. H. 1975 *Statistical energy analysis of dynamical systems: theory and applications*. MIT Press.

Ming, R. S., Stimpson, G. & Lalor, N. 1990 On the measurement of individual coupling loss factors in a complex-structure. In *Inter-Noise 90. Poughkeepsie: U.S.A.: Noise Control Foundation.*

Stewart, G. W. 1973 *Introduction to matrix computations*. New York: Academic Press.

Westphal, W. 1957 Ausbreitung von Körperschall in Gebäuden. *Acustica* **7**, 335–348.

Woodhouse, J. 1981 An introduction to statistical energy analysis applied to structural vibrations. *Appl. Acoustics* **14**, 444–469.

Measurement of SEA coupling loss factors using point mobilities

By C. Cacciolati and J. L. Guyader

Laboratoire Vibrations Acoustique, Institut National des Sciences Appliquées, Bat 303, 20 Avenue Albert Einstein, 69100 Villeurbanne, France

In this paper calculation of SEA coupling loss factors, using measured point mobility, is derived for coupled systems, homogeneous or not, with rigid or soft links. Some simplifications and hypothesis are necessary to fit with SEA basic relations. To validate the theory an experiment was done on plates coupled in three points; the agreement is reasonably good for homogeneous and non-homogeneous plates.

1. Introduction

The SEA method (see Lyon & Maidanik 1962; Scharton & Lyon 1968) offer a good tool to analyse and predict acoustic and vibration transmissions in coupled systems (see Craik (1982) for building structures and Plunt (1980) for ships). From the practical point of view, the capital element for applying SEA is the determination of the coupling loss factors (CLF). Theoretical estimations of CLF can be obtained in simple cases from wave propagation in infinite coupled beams, plates, shells (for plates see Gibbs & Guilford 1976, 1987; Wohle *et al.* 1981; Van Backel & De Vries 1983). In general cases Keane & Price (1987) gave a relation for CLF using direct and cross receptance of coupled systems. It is also possible to derive the CLF from Green's function of decoupled structure, as demonstrated by Davies & Wahab (1981) on two coupled beams.

When people deal with real industrial structures, difficulties arise; the structures are non-academic due to complicated shapes and heterogeneities, the links are often non-rigid, dissipative and localized at some points. In this case it is impossible to calculate theoretically the coupling loss factors, except if a finite element modelization is used (see Simmons (1991) for example). However, this technique is not easy to use in the acoustic frequency range. A second possibility consists of measuring the CLF, by an experimental identification (see Lalor 1987). It is based on an inverse SEA procedure, which calculates the CLF, from the energies of the coupled systems. Two difficulties can be noticed; the systems must be coupled to apply the method, thus it is impossible to predict CLF from measurement on decoupled structures. The calculation of CLF from energies necessitates the resolution of a linear system that can be ill conditioned, this is related to the relative insensibility of SEA energy prediction, to modifications of coupling loss factors.

The study of structures coupled by point links, can be made with the mobility technique; one has to characterize the coupling points with their direct and transfer mobilities as presented, for example, by Hemingway (1986). On complicated structures (e.g. industrial) the mobilities must be measured. These measurements, contrary to inverse SEA procedure, are done on decoupled structures. The difficulties in applying this method are because of the appearance of singular frequencies, due to the cumulation of errors, in particular when the number of links is large. The precise measurement of mobilities is thus very important for this technique, unfortunately the uncertainty is large, especially on the phase and when frequency is high. It seems interesting to us to mix the two approaches to calculate CLF with measured mobilities. This is interesting from two points of view: (i) the CLF are determined from measurements on decoupled structures; and (ii) only the active power is necessary. The reactive part, difficult to measure, can be ignored.

2. Direct and transfer mobilities

Let us consider two systems coupled in some points, and excited at a given angular frequency ω. The problem consists in calculating the coupled response, using decoupled behaviour, of each system. This can be done by introducing mobilities $M_{\mathrm{I}}(Q_i, Q_j)$ defined as the velocity of system I at point Q_i when point Q_j is excited by a force of unit amplitude and angular frequency ω. It is a complex quantity depending on frequency. If the two systems are excited they have a velocity before coupling of $\tilde{V}_{\mathrm{I}}(Q_i)$ and $\tilde{V}_{\mathrm{II}}(Q'_j)$, the velocity of each structure after coupling can be obtained at each frequency with relations (1) and (2), based on linearity of the systems and links:

$$V_{\mathrm{I}}(Q_i) = \tilde{V}_{\mathrm{I}}(Q_i) + \sum_{j=1}^{N} M_{\mathrm{I}}(Q_i, \bar{Q}_j)\, F^{\mathrm{c}}_{\mathrm{I}}(\bar{Q}_j), \tag{1}$$

$$V_{\mathrm{II}}(Q'_i) = \tilde{V}_{\mathrm{II}}(Q'_i) + \sum_{j=1}^{N} M_{\mathrm{II}}(Q'_i, \bar{Q}'_j)\, F^{\mathrm{c}}_{\mathrm{II}}(\bar{Q}'_j), \tag{2}$$

where an overbar indicates a coupling point, and a prime a point of system II. $\bar{Q}_i$ and $\bar{Q}'_j$ are the coupling points respectively on system I and II. The coupling forces acting on system I and II are $F^{\mathrm{c}}_{\mathrm{I}}(\bar{Q}_j)$, $F^{\mathrm{c}}_{\mathrm{II}}(\bar{Q}'_j)$. To calculate the coupling forces, in the case of rigid links, one has to write equality of velocity and equilibrium of forces, at the coupling points:

$$V_{\mathrm{I}}(Q_i) = V_{\mathrm{II}}(Q'_i), \tag{3}$$

$$F^{\mathrm{c}}_{\mathrm{I}}(\bar{Q}_j) + F^{\mathrm{c}}_{\mathrm{II}}(\bar{Q}'_j) = 0. \tag{4}$$

After calculation one obtains

$$\{F^{\mathrm{c}}_{\mathrm{I}}(\bar{Q}_i)\} = [M_{\mathrm{I}}(\bar{Q}_i, \bar{Q}_j) + M_{\mathrm{II}}(\bar{Q}'_i, \bar{Q}'_j)]^{-1}\{\tilde{V}_{\mathrm{II}}(\bar{Q}'_j) - \tilde{V}_{\mathrm{I}}(\bar{Q}_j)\}. \tag{5}$$

3. The case of one rigid link

To simplify the analysis, let us consider only one link, equation (5) reduces to

$$F^{\mathrm{c}}_{\mathrm{I}}(\bar{Q}_1) = \frac{\tilde{V}_{\mathrm{II}}(\bar{Q}'_1) - \tilde{V}_{\mathrm{I}}(\bar{Q}_1)}{M_{\mathrm{I}}(\bar{Q}_1, \bar{Q}_1) + M_{\mathrm{II}}(\bar{Q}'_1, \bar{Q}'_1)}. \tag{6}$$

It is now easy to calculate the power injected in system I and II:

$$\Pi^{\mathrm{I}}(\bar{Q}_1) = \tfrac{1}{2}\mathrm{Re}\{F^{\mathrm{c}}_{\mathrm{I}}(\bar{Q}_1)\, V^*_{\mathrm{I}}(\bar{Q}_1)\}, \tag{7}$$

$$\Pi^{\mathrm{II}}(\bar{Q}'_1) = \tfrac{1}{2}\mathrm{Re}\{F^{\mathrm{c}}_{\mathrm{II}}(\bar{Q}'_1)\, V^*_{\mathrm{II}}(\bar{Q}'_1)\}. \tag{8}$$

The rigid link being non dissipative, the injected power in each systems are opposite. The power going out of system II through the link is equal to the injected power in system I:

$$\Pi_{\mathrm{II}\to\mathrm{I}} = \tfrac{1}{4}\mathrm{Re}\{(\tilde{V}_I + \tilde{V}_{\mathrm{II}})F^* + (M_{\mathrm{I}} - M_{\mathrm{II}})FF^*\}. \tag{9}$$

To obtain (9) one has to use (7) and (8) and simplify the notation

$$\tilde{V}_{\mathrm{I}} = \tilde{V}_{\mathrm{I}}(\bar{Q}_1), \quad M_{\mathrm{I}} = M_{\mathrm{I}}(\bar{Q}_1, \bar{Q}_1), \quad \tilde{V}_{\mathrm{II}} = \tilde{V}_{\mathrm{II}}(\bar{Q}'_1), \quad M_{\mathrm{II}} = M_{\mathrm{II}}(\bar{Q}'_1, \bar{Q}'_1),$$
$$F^{\mathrm{c}}_{\mathrm{I}}(\bar{Q}_1) = -F^{\mathrm{c}}_{\mathrm{II}}(\bar{Q}'_1) = F.$$

Replacing F with expression (6) gives, after calculations,

$$\Pi_{\mathrm{II}\to\mathrm{I}} = \frac{1}{2}\left[|\tilde{V}_{\mathrm{II}}|^2 \frac{\mathrm{Re}\,(M_{\mathrm{I}})}{|M_{\mathrm{I}} + M_{\mathrm{II}}|^2} - |\tilde{V}_{\mathrm{I}}|^2 \frac{\mathrm{Re}\,(M_{\mathrm{II}})}{|M_{\mathrm{I}} + M_{\mathrm{II}}|^2}\right] + \frac{1}{2}\left[\frac{\mathrm{Re}\,(\tilde{V}_{\mathrm{I}}\tilde{V}^*_{\mathrm{II}} M_{\mathrm{II}} - \tilde{V}^*_{\mathrm{I}}\tilde{V}_{\mathrm{II}} M_{\mathrm{I}})}{|M_{\mathrm{I}} + M_{\mathrm{II}}|^2}\right]. \tag{10}$$

This equation is valid for a single frequency analysis, in the case of band excitation it must be integrated over frequency. When non-correlated vibrations are considered the third term of (10) integrated over frequency is neglected. Equation (10) reduces to

$$\langle \Pi_{\mathrm{II}\to\mathrm{I}} \rangle = \frac{1}{2}\left\langle |\tilde{V}_{\mathrm{II}}|^2 \frac{\mathrm{Re}\,(M_{\mathrm{I}})}{|M_{\mathrm{I}} + M_{\mathrm{II}}|^2}\right\rangle - \frac{1}{2}\left\langle |\tilde{V}_{\mathrm{I}}|^2 \frac{\mathrm{Re}\,(M_{\mathrm{II}})}{|M_{\mathrm{I}} + M_{\mathrm{II}}|^2}\right\rangle, \tag{11}$$

where brackets signify integration over a frequency band $\Delta\omega$.

If in addition one supposes that decoupled vibration fields vary smoothly with frequency, it is possible to approximate (11) with

$$\langle \Pi_{\mathrm{II}\to\mathrm{I}} \rangle = \frac{1}{2}\frac{\langle |\tilde{V}_{\mathrm{II}}|^2\rangle}{\Delta\omega}\left\langle \frac{\mathrm{Re}\,(M_{\mathrm{I}})}{|M_{\mathrm{I}} + M_{\mathrm{II}}|^2}\right\rangle - \frac{1}{2}\frac{\langle |\tilde{V}_{\mathrm{I}}|^2\rangle}{\Delta\omega}\left\langle \frac{\mathrm{Re}\,(M_{\mathrm{II}})}{|M_{\mathrm{I}} + M_{\mathrm{II}}|^2}\right\rangle. \tag{12}$$

Physically this assumption corresponds to systems having sufficiently high modal overlap. This equation relates the power flow from system II to system I, with local velocities of both systems before coupling.

To introduce energies one has to assume that the vibration fields are homogeneous and so vary slowly with location. In homogeneous systems it was demonstrated by Dowell & Kubota (1985), except near the boundary. The equality of kinetic and deformation energies is assumed, thus for decoupled system, the total energy over the bandwidth $\Delta\omega$ is

$$\langle \tilde{E}_{\mathrm{I}} \rangle = \tfrac{1}{2}\langle [\langle\langle \mu_{\mathrm{I}} |\tilde{V}_{\mathrm{I}}|^2 \rangle\rangle] \rangle \approx \tfrac{1}{2}\langle\langle \mu_{\mathrm{I}} \rangle\rangle \langle |\tilde{V}_{\mathrm{I}}|^2 \rangle, \tag{13}$$

where the density of mass of system I is μ_{I} and the double bracket represents spacial integration over the system. The following can be applied if the coupling point is not situated close to boundary. By using (12) in connection with (13) gives

$$\langle \Pi_{\mathrm{II}\to\mathrm{I}} \rangle = [\eta_{\mathrm{II\,I}} \langle \tilde{E}_{\mathrm{II}} \rangle - \eta_{\mathrm{I\,II}} \langle \tilde{E}_{\mathrm{I}} \rangle]\,\omega, \tag{14}$$

with

$$\eta_{\mathrm{II\,I}} = \frac{1}{m_{\mathrm{II}}\,\omega}\left[\frac{1}{\Delta\omega}\left\langle \frac{\mathrm{Re}\,(M_{\mathrm{I}})}{|M_{\mathrm{I}} + M_{\mathrm{II}}|^2}\right\rangle\right], \tag{15}$$

$$\eta_{\mathrm{I\,II}} = \frac{1}{m_{\mathrm{I}}\,\omega}\left[\frac{1}{\Delta\omega}\left\langle \frac{\mathrm{Re}\,(M_{\mathrm{II}})}{|M_{\mathrm{I}} + M_{\mathrm{II}}|^2}\right\rangle\right], \tag{16}$$

where $m_{\mathrm{I}} = \langle\langle \mu_{\mathrm{I}} \rangle\rangle$ is the total mass of the system and ω is the centre band frequency.

A last assumption is necessary to identify with SEA relation, it is the weak coupling hypothesis in the sense of energies varying weakly before and after coupling. Let us introduce the energies after coupling E_{I} and E_{II}:

$$\langle E_{\mathrm{I}}\rangle = \langle \tilde{E}_{\mathrm{I}}\rangle + \langle e_{\mathrm{I}}\rangle \quad \text{and} \quad \langle E_{\mathrm{II}}\rangle = \langle \tilde{E}_{\mathrm{II}}\rangle + \langle e_{\mathrm{II}}\rangle. \tag{17}$$

Equation (14) gives

$$\langle \Pi_{\mathrm{II}\to\mathrm{I}}\rangle = \omega[\eta_{\mathrm{II\,I}}\langle E_{\mathrm{II}}\rangle - \eta_{\mathrm{I\,II}}\langle E_{\mathrm{I}}\rangle - \eta_{\mathrm{II\,I}}\langle e_{\mathrm{II}}\rangle + \eta_{\mathrm{I\,II}}\langle e_{\mathrm{I}}\rangle], \tag{18}$$

and the weak coupling assumes that one can neglect the two last terms in (18).

$$\left.\begin{aligned}
\langle \Pi_{\mathrm{II}\to\mathrm{I}}\rangle &\approx \omega[\eta_{\mathrm{II\,I}}\langle E_{\mathrm{II}}\rangle - \eta_{\mathrm{I\,II}}\langle E_{\mathrm{I}}\rangle],\\
\eta_{\mathrm{II\,I}} &= \frac{1}{m_{\mathrm{II}\omega}}\left[\frac{1}{\Delta\omega}\left\langle\frac{\mathrm{Re}\,(M_{\mathrm{I}})}{|M_{\mathrm{I}}+M_{\mathrm{II}}|^2}\right\rangle\right],\\
\eta_{\mathrm{I\,II}} &= \frac{1}{m_{\mathrm{I}\,\omega}}\left[\frac{1}{\Delta\omega}\left\langle\frac{\mathrm{Re}\,(M_{\mathrm{II}})}{|M_{\mathrm{I}}+M_{\mathrm{II}}|^2}\right\rangle\right].
\end{aligned}\right\} \tag{19}$$

This equation is the basic SEA relation, it shows that coupling loss factors can be calculated from point mobilities with (19).

4. The case of several rigid links

Equation (19) is established for one rigid link. In a practical situation several links must often be considered and an extension to this case is of interest. When several links are coupling the two systems, the exact calculation of powers exchanged is cumbersome, as transfer mobilities must be used. From the experimental point of view, measurement of direct and transfer mobilities is not easy when the number of links increases. So an important question arises: are transfer mobilities strongly modifying the transmitted power from one system to the other? In other words is each link exchanging power independently from the others? To answer this question we consider a system of two plates coupled by three rigid links. The geometry of the system is the same as the one used in experiment (see figure 2) but plates here are simply supported to simplify the calculation. The kinetic energy T of the receiver plate II is calculated when a unit force is applied at point E on the plate I. Figure 1 presents the difference in dB between the exact solution T_{e} and approximation neglecting transfer mobilities T_{a}. For systems of high modal density and when an average over frequency is done, the influence of transfer mobility is negligible.

In consequence coupling loss factors for several rigid links can be obtained by summation of expressions (19) over points of coupling

$$\eta_{\mathrm{II\,I}} = \frac{1}{m_{\mathrm{II}}\,\omega}\sum_{k=1}^{N}\frac{1}{\Delta\omega}\left\langle\frac{\mathrm{Re}\,(M_{\mathrm{I}}^{k})}{|M_{\mathrm{I}}^{k}+M_{\mathrm{II}}^{k}|^2}\right\rangle, \tag{20}$$

$$\eta_{\mathrm{I\,II}} = \frac{1}{m_{\mathrm{I}}\,\omega}\sum_{k=1}^{N}\frac{1}{\Delta\omega}\left\langle\frac{\mathrm{Re}\,(M_{\mathrm{II}}^{k})}{|M_{\mathrm{I}}^{k}+M_{\mathrm{II}}^{k}|^2}\right\rangle. \tag{21}$$

The exponent k indicates the link number, and N is the total number of coupling points. For large values of N, where the coupling points become close to each other

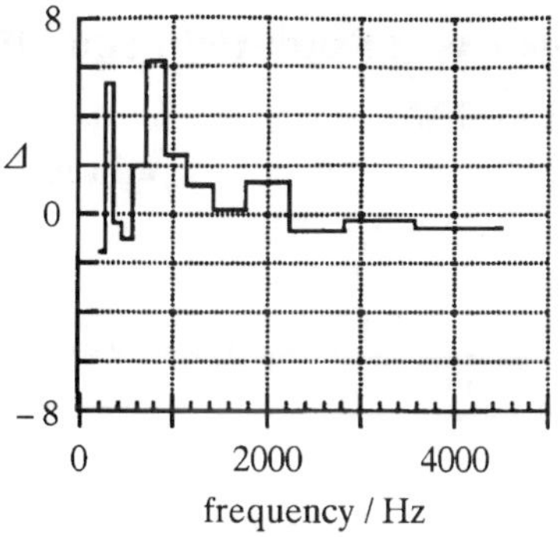

Figure 1. Influence of the transfer mobility: $\Delta = 10\log(T_e) - 10\log(T_a)$ against frequency in Hertz.

it is important to verify that the transfer mobilities can be still neglected. In other words links must be sufficiently distant otherwise it is necessary to use a link or surface effective mobility (see, for example, Naji *et al.* 1992; Petersson & Plunt 1982).

5. The case of non-homogeneous structures

In the previous analysis, vibration fields are supposed homogeneous, for industrial structures this assumption is not true in general. If, as previously assumed, decoupled vibrations are non-correlated, and structures are coupled rigidly in one point, equation (12) remains valid. The difference between homogeneous and non-homogeneous systems comes from equation (13) as local quadratic velocity can vary strongly in non-homogeneous systems. Let us introduce the homogeneity factor g calculated on the decoupled structure with their own excitation, over the bandwidth frequency $\Delta\omega$. It is the non-dimensional ratio indicated by (22). Physically it corresponds to the ratio of coupling point energy with the total mass of the system over total energy of the system.

$$g_{\mathrm{I}} = \frac{\langle|\tilde{V}_{\mathrm{I}}(\bar{Q}_1)|^2\rangle\, m_{\mathrm{I}}}{2\langle\tilde{E}_{\mathrm{I}}\rangle} \quad \text{and} \quad g_{\mathrm{II}} = \frac{\langle|\tilde{V}_{\mathrm{II}}(\bar{Q}'_1)|^2\rangle\, m_{\mathrm{II}}}{2\langle\tilde{E}_{\mathrm{II}}\rangle}. \tag{22}$$

This homogeneity factor is equal to 1 for homogeneous vibration field. The equation (14) can then be used but with modified coupling loss factors

$$\left.\begin{aligned} \eta_{\mathrm{II\,I}} &= \frac{g_{\mathrm{II}}}{m_{\mathrm{II}}\,\omega}\left[\frac{1}{\Delta\omega}\left\langle\frac{\mathrm{Re}\,(M_{\mathrm{I}})}{|M_{\mathrm{I}}+M_{\mathrm{II}}|^2}\right\rangle\right], \\ \eta_{\mathrm{I\,II}} &= \frac{g_{\mathrm{I}}}{m_{\mathrm{I}}\,\omega}\left[\frac{1}{\Delta\omega}\left\langle\frac{\mathrm{Re}\,(M_{\mathrm{II}})}{|M_{\mathrm{I}}+M_{\mathrm{II}}|^2}\right\rangle\right]. \end{aligned}\right\} \tag{23}$$

For several coupling points which verify the assumptions of §4, equations (13) and (14) remain valid. It is only necessary to sum the power flow of each link (see (21) and (22)),

$$\left.\begin{aligned} \eta_{\mathrm{II\,I}} &= \frac{1}{m_{\mathrm{II}}\,\omega}\sum_{k=1}^{N}\left[g^k_{\mathrm{II}}\frac{1}{\Delta\omega}\left\langle\frac{\mathrm{Re}\,(M^k_{\mathrm{I}})}{|M^k_{\mathrm{I}}+M^k_{\mathrm{II}}|^2}\right\rangle\right], \\ \eta_{\mathrm{I\,II}} &= \frac{1}{m_{\mathrm{I}}\,\omega}\sum_{k=1}^{N}\left[g^k_{\mathrm{I}}\frac{1}{\Delta\omega}\left\langle\frac{\mathrm{Re}\,(M^k_{\mathrm{II}})}{|M^k_{\mathrm{I}}+M^k_{\mathrm{II}}|^2}\right\rangle\right]. \end{aligned}\right\} \tag{24}$$

6. The case of non-rigid coupling

When dealing with industrial links, one has often to consider non-rigid coupling, in this case equations (3) and (4) are no more true and must be replaced by a matrix relation between point velocities after coupling, and coupling forces acting on both systems (F_{I}) and F_{II})

$$\begin{bmatrix} V_{\mathrm{I}} \\ V_{\mathrm{II}} \end{bmatrix} = \begin{bmatrix} A_{\mathrm{I\,I}} & A_{\mathrm{I\,II}} \\ A_{\mathrm{II\,I}} & A_{\mathrm{II\,II}} \end{bmatrix} \begin{bmatrix} F_{\mathrm{I}} \\ F_{\mathrm{II}} \end{bmatrix}. \tag{25}$$

The terms of the matrix characterize mobilities of the link; $A_{\mathrm{I\,I}}$ (resp. $A_{\mathrm{II\,II}}$) is the direct mobility at coupling point with system I (resp. II), $A_{\mathrm{I\,II}}$ and $A_{\mathrm{II\,I}}$ are transfer mobilities. The coupling forces can be calculated equating the velocity after coupling of the link and of the structure, one obtains the coupling forces from decoupled velocities of structures at coupling points:

$$\begin{bmatrix} F_{\mathrm{I}} \\ F_{\mathrm{II}} \end{bmatrix} = \frac{1}{\Delta} \begin{bmatrix} A_{\mathrm{II\,II}} - M_{\mathrm{II}} & -A_{\mathrm{I\,II}} \\ -A_{\mathrm{II\,I}} & A_{\mathrm{I\,I}} - M_{\mathrm{I}} \end{bmatrix} \begin{bmatrix} \tilde{V}_{\mathrm{I}} \\ \tilde{V}_{\mathrm{II}} \end{bmatrix}, \tag{26}$$

where $\Delta = (A_{\mathrm{I\,I}} - M_{\mathrm{I}})(A_{\mathrm{II\,II}} - M_{\mathrm{II}}) - A_{\mathrm{I\,II}} A_{\mathrm{II\,I}}$. It is now easy to calculate powers injected in each system

$$\Pi^{\mathrm{I}} = \tfrac{1}{2} \mathrm{Re}\,[F_{\mathrm{I}}^{*} V_{\mathrm{I}}] \quad \text{and} \quad \Pi^{\mathrm{II}} = \tfrac{1}{2} \mathrm{Re}\,[F_{\mathrm{II}}^{*} V_{\mathrm{II}}], \tag{27}$$

only, of course, if the coupling is dissipative. The powers Π^{I} and Π^{II} are not opposite, and classical SEA relation is no more applicable, on the contrary for non-dissipative coupling, assuming non-correlated decoupled vibrations and smooth variation with frequency of decoupled quadratic velocities, one obtains instead of (12)

$$\Pi_{\mathrm{I}\to\mathrm{II}} = \tfrac{1}{4} |\tilde{V}_{\mathrm{II}}|^2 \left\langle \frac{|A_{\mathrm{I\,II}}|^2 - \mathrm{Re}\,[(\Delta + M_{\mathrm{II}} A_{\mathrm{I\,I}} - M_{\mathrm{I}} M_{\mathrm{II}})(A_{\mathrm{I\,I}}^{*} - M_{\mathrm{I}}^{*})]}{|\Delta|^2} \right\rangle$$

$$- \tfrac{1}{4} |\tilde{V}_{\mathrm{I}}|^2 \left\langle \frac{|A_{\mathrm{II\,I}}|^2 - \mathrm{Re}\,[(\Delta + M_{\mathrm{I}} A_{\mathrm{II\,II}} - M_{\mathrm{I}} M_{\mathrm{II}})(A_{\mathrm{II\,II}}^{*} - M_{\mathrm{II}}^{*})]}{|\Delta|^2} \right\rangle. \tag{28}$$

The coupling loss factors can then be calculated as previously, one obtains for non-homogeneous structures

$$\eta_{\mathrm{II\,I}} = \frac{1}{m_{\mathrm{II}}\,\omega} \frac{1}{\Delta\omega} g_{\mathrm{II}} \left\langle \frac{|A_{\mathrm{I\,II}}|^2 - \mathrm{Re}\,[(\Delta + M_{\mathrm{II}} A_{\mathrm{I\,I}} - M_{\mathrm{I}} M_{\mathrm{II}})(A_{\mathrm{I\,I}}^{*} - M_{\mathrm{I}}^{*})]}{2|\Delta|^2} \right\rangle, \tag{29}$$

$$\eta_{\mathrm{I\,II}} = \frac{1}{m_{\mathrm{I}}\,\omega} \frac{1}{\Delta\omega} g_{\mathrm{I}} \left\langle \frac{|A_{\mathrm{II\,I}}|^2 - \mathrm{Re}\,[(\Delta + M_{\mathrm{I}} A_{\mathrm{II\,II}} - M_{\mathrm{I}} M_{\mathrm{II}})(A_{\mathrm{II\,II}}^{*} - M_{\mathrm{II}}^{*})]}{2|\Delta|^2} \right\rangle. \tag{30}$$

7. Experimental validation

The different expressions of the coupling loss factors previously presented are established with several assumptions and simplifications. Their validity must then be demonstrated on experimental ground, several experiments were done on coupled plates to calculate coupling loss factors from measured mobilities and then to compare energy of plates using SEA relations and direct calculations.

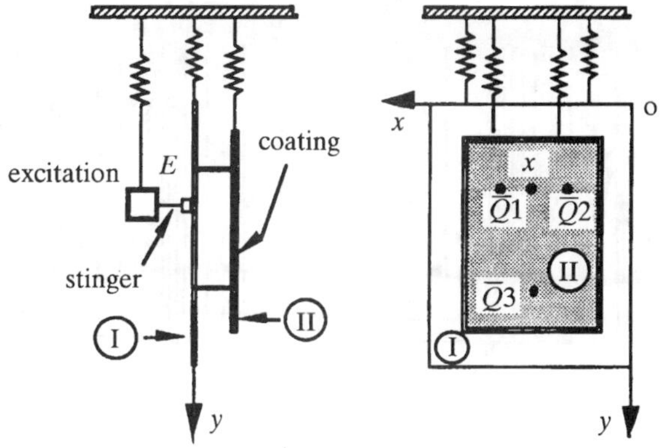

Figure 2. Experiment.

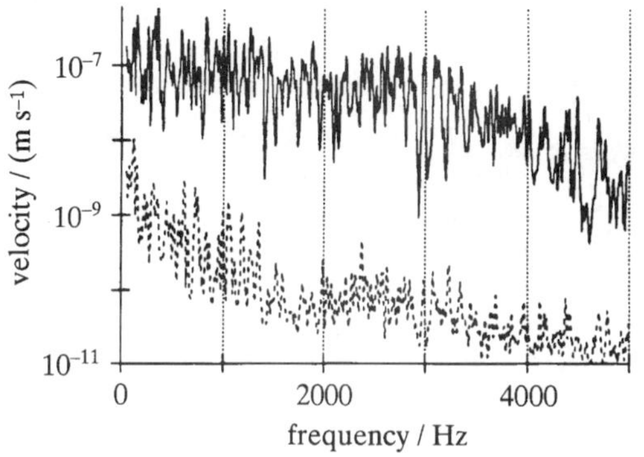

Figure 3. Plates velocities (m s^{-1}) against frequency, without mechanical links. ——, Point X_{I} on excited plate; ······, point X_{II} on receiving plate.

(*a*) *The case of homogeneous plates*

Two plates were coupled with three mechanical links to connect rigidly transverse displacement but not flexural rotation, this was achieved using stingers (see figure 2). To permit power flow, plate II was covered with a layer of viscoelastic material. Plates were made of steel, they had free edges and were attached by soft springs on a rigid frame. Plate I was $0.7 \text{ m} \times 1.0 \text{ m} \times 0.002 \text{ m}$ and plate II was $0.5 \text{ m} \times 0.75 \text{ m} \times 0.002 \text{ m}$. The air gap between plates was 0.085 m. The coordinates of the coupling points $\bar{Q}_i$ and driving point E were given on the general coordinate system X, Y: $E(0.22 \text{ m}, 0.36 \text{ m})$, $\bar{Q}_1(0.5 \text{ m}, 0.23 \text{ m})$, $\bar{Q}_2(0.195 \text{ m}, 0.23 \text{ m})$, $\bar{Q}_3(0.35 \text{ m}, 0.77 \text{ m})$. The masses were 10.35 kg for plate I, 6.40 kg for plate II, 0.0051 kg for each link, 0.003 kg for the dynamic added mass by the force transducer and its screw of fixation. The first resonance frequency of links considered simply supported at each end was 4150 Hz.

Figure 3 shows an experimental result on the plates velocities, when the mechanical links are removed to determine if acoustic transmission is negligible compared with mechanical transmission. The direct excitation is applied at the point E on plate I, and the plate II is excited by the acoustical transmission through the air gap. The ratio of the speeds measured at points X_{I} (resp. X_{II}) located on the middle of plate I (resp. II) is greater than 100 instead of 2 with mechanical links, so the acoustical transmission through the air gap between the plates may be neglected.

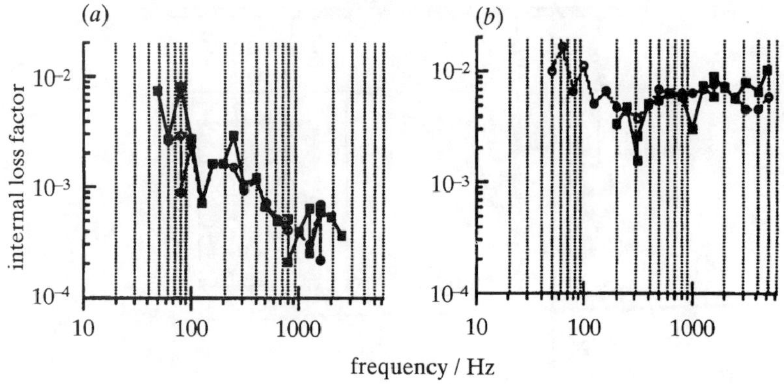

Figure 4. Internal loss factor against frequency in Hertz. (a) Plate I, (b) plate II. —■—, pink noise excitation; —□—, hammer shock excitation.

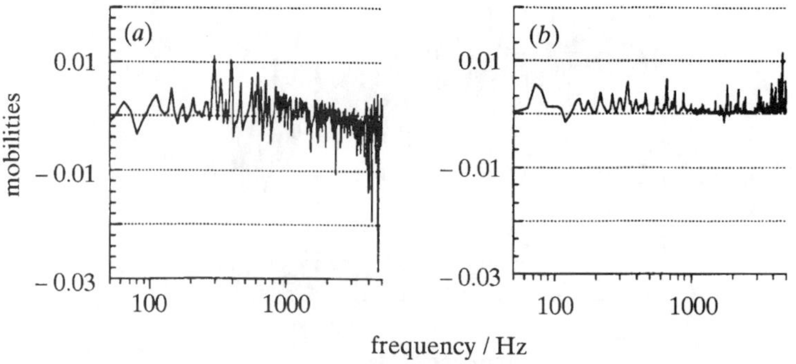

Figure 5. Point mobilities on the receiving plate, at coupling point Q3 against frequency in Hertz. (a) Imaginary part, (b) real part.

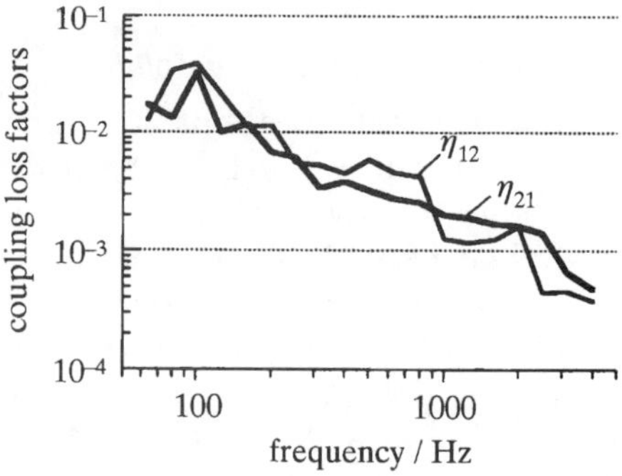

Figure 6. Calculated coupling loss factors expressions (20) and (21) against frequency in $\frac{1}{3}$ octave bandwidth.

Measurements of internal loss factors against frequency of decoupled plates are presented in figure 4. They were deduced from a one-third octave measurement of the reverberation time.

The mobilities of decoupled plates at the junction points were measured, figure 5 shows a typical result. The values of the mobilities do not vary strongly with point of measurement, this is consistent with the homogeneity of the plate. Applying expressions (20) and (21) of the coupling loss factors one get the results shown in figure 6.

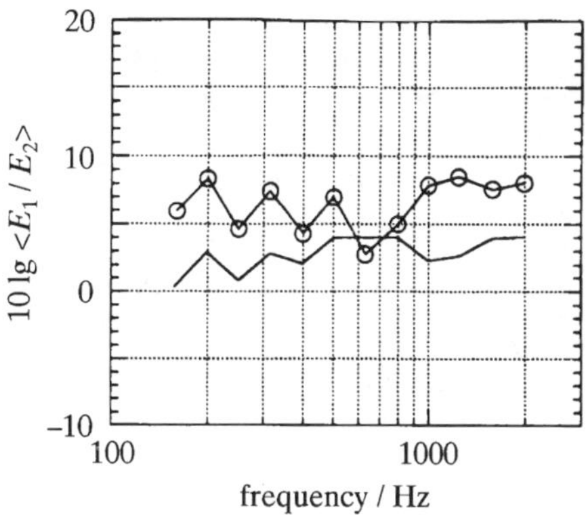

Figure 7. $10\lg(\langle E_1\rangle/\langle E_2\rangle)$ against frequency in Hertz. ——, SEA calculation; ‒‒○‒‒, direct measurement.

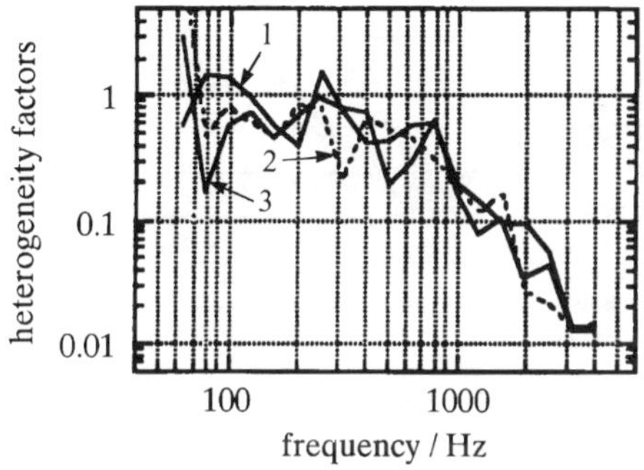

Figure 8. Heterogeneity factors against frequency in Hertz. 1, $\bar{Q}_1$; 2, point $\bar{Q}_2$; 3, $\bar{Q}_3$.

The two CLF follow the same tendency, they decrease with frequency from 0.01 to 0.001, and remain close together in the whole frequency range.

To verify the validity of calculated coupling loss factors an experiment was done on the two coupled plates described previously. Figure 7 compares the measured plate energy ratio $\langle E_1\rangle/\langle E_2\rangle$ and the SEA prediction obtained, introducing loss factors presented in figure 6, in equation (31):

$$\langle E_1\rangle/\langle E_2\rangle = (\eta_{21}+\eta_2)/\eta_{12}. \tag{31}$$

The SEA prediction overestimates the transmission in general, but the difference between the two curves is reasonable in average. One also notices the variation of the damping loss factor of the receiving plate, in equation (31), introduces strong variation of the SEA energy ratio that can explain the overestimation.

(*b*) *The case of non-homogeneous plate*

To check the validity of expressions (24) of coupling loss factors for non-homogeneous structures, the experiment described in §7*a* was modified. Three added masses (0.175 kg each) were placed on the excited plate at points of coupling.

Figure 8 presents the measured heterogeneity factors associated with each added mass. The three curves are close together and decrease with frequency. This tendency is consistent with the added mass behaviour that blocks the movement when the frequency increases.

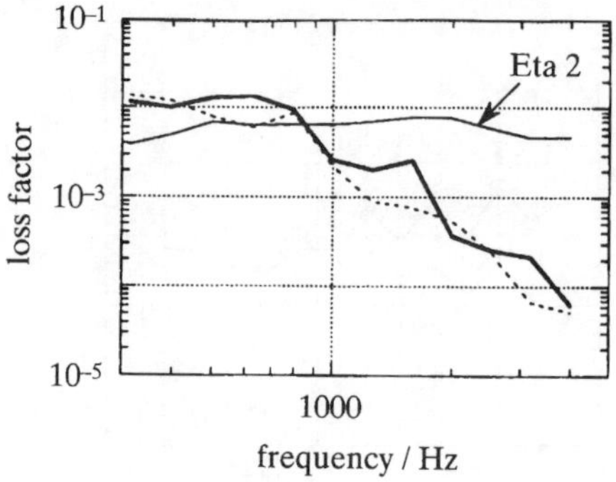

Figure 9. Internal loss factor of structure 2 and SEA coupling loss factors η_{12} (– – – –) and η_{21} (——) against frequency.

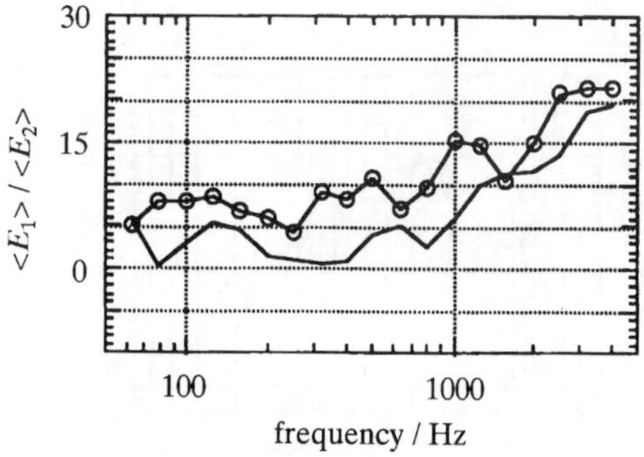

Figure 10. Ratio $\langle E_1\rangle/\langle E_2\rangle$ in dB against frequency in Hertz, where $\mathrm{dB} = 10\log(\langle E_1\rangle/\langle E_2\rangle)$. – –○– –, Direct measurement; ——, SEA prediction.

The coupling loss factors were calculated according to (24) and the results are presented in figure 9. The coupling loss factors have close values and decrease strongly with frequency (compared with the experiment on homogeneous plates, figure 6). In the high-frequency range, the internal damping of the receiver structure is greater than coupling loss factors and thus governs the transmission. The ratio of the plate energies is close to 2 below 1000 Hz, and increases after, up to 130 at 4000 Hz (see figure 10). The predicted value using SEA coupling loss factors overestimate the transmission as for homogeneous plates, but give the right tendency versus frequency.

8. Conclusion

Determination of SEA coupling loss factors from measured mobilities on decoupled structures have been presented. In the case of two plates coupled in three points, the validity of the expression derived for coupling loss factors, has been demonstrated from an experiment. The heterogeneity factor must be introduced for heterogeneous structure, and used as a weighting function in CLF expression.

The derivation of SEA relations from mobility concepts allows us to introduce non-rigid coupling characterized by input and transfer mobility. The validity of the approach will be checked on experimental ground in the future.

The results given here are applicable for weakly coupled subsystems with high modal overlap. More complex relations would be needed in other circumstances.

References

Craik, R. J. M. 1982 The prediction of sound transmission on building using S.E.A. *J. Sound Vib.* **82**, 505–516.

Davies, H. G. & Wahab, M. A. 1981 Ensemble averages of power flow in randomly excited coupled beams. *J. Sound Vib.* **77**, 311–321.

Dowell, E. H. & Kubuta, Y. 1985 Asymptotic modal analysis and statistical energy analysis of dynamical systems. *J. appl. Mech.* **52**, 949–957.

Gibbs, B. M. & Guilford, C. L. 1976 The use of power flow methods for the assessment of sound transmission in building structures. *J. Sound. Vib.* **49**, 267–286.

Gibbs, B. M. & Guilford, C. L. 1987 Prediction by power flow methods of shunt and series damping in building structures. *Appl. Acoust.* **10**, 291–301.

Hemingway, N. G. 1986 Modelling vibration transmission through coupled vehicle sub system using mobility matrix. *J. Mech. Engng* **200**, 125–135.

Keane, A. J. & Price, W. G. 1987 Statistical energy analysis of strongly coupled systems. *J. Sound Vib.* **117**, 363–386.

Lalor, N. 1987 The measurement of S.E.A. loss factor on a fully assembled structure. I.S.V.R. technical report no. 150.

Lyon, R. H. & Maidanik, G. 1962 Power flow between linearly coupled oscillators. *J. acoust. Soc. Am.* **34**, 623–639.

Naji, S., Cacciolati, C. & Guyader, J. L. 1992 Modèles de transmissions vibratoire par les couplages surfaciques. *Colloque Cl. Sup. J. Phys.* **III**, 519–522.

Petersson, B. & Plunt, J. 1982 On effective mobilities, in the prediction of structure borne sound transmission between a source structure and a receiver structure. *J. Sound Vib.* **82**, part I SA-529, part II 531–540.

Plunt, J. 1980 Methods for predicting noise levels in ships. Experiences from empirical and S.E.A. calculation methods. Report Chalmers University, Goteborg, Sweden.

Scharton, T. D. & Lyon, R. H. 1968 Power flow and energy sharing in random vibrations. *J. acoust. Soc. Am.* **43**, 1332–1343.

Simmons, C. 1991 Structure borne sound transmission through plates junctions and estimates of S.E.A. coupling loss factors using the finite element model. *J. Sound Vib.* **144**, 215–221.

Van Bakel, J. G. & De Vries, D. 1983 Parameter sensitivity of a T junction S.E.A. model, the importance of the internal damping loss factor. *J. Sound Vib.* **90**, 373–380.

Wöhle, W., Beckmann, T. & Schreckenbach, H. 1981 Coupling loss factors for S.E.A. of sound transmission at rectangular slab joints. *J. Sound Vib.* **77**, Part I 323–334, Part II 335–344.

References

Formulation of SEA parameters using mobility functions

By Jerome E. Manning
Cambridge Collaborative, Inc., 689 Concord Avenue, Cambridge, Massachusetts 02138, U.S.A.

Statistical energy analysis (SEA) formulates the dynamic response of a system in terms of power and energy variables. The SEA parameters include power inputs; damping loss factors; which control the power dissipated within the system; and coupling loss factors, which control the power transmitted between coupled subsystems. One of the great difficulties in using SEA is the calculation of these parameters. In this paper SEA parameters are formulated using general mobility functions. Simplifications that result from averaging the parameters either over frequency or over an ensemble of dynamic systems are presented. These simplifications make it possible to apply SEA to very complex structural-acoustic systems.

1. Introduction

Statistical energy analysis, or SEA as it is commonly called, provides a technique to study the dynamic response of complex structural-acoustic systems. Since its introduction almost 30 years ago, SEA has slowly grown in popularity. Today it is considered by many to be the best analysis technique for high frequencies (Rockwood 1987). In addition, research is being conducted at Cambridge Collaborative and other organizations to extend the validity of the technique to lower frequencies (Manning 1990). The rise in popularity of SEA is due to the fact that it can be used both as a theoretical technique and an experimental technique. The theoretician computes the SEA parameters using theoretical mode and wave analysis, while the experimentalist computes these parameters using various measured data. While both techniques have merit, this paper focuses more on the theoretical side of SEA than the experimental. It is hoped that those trying to compute SEA parameters from basic theory will benefit from the approaches outlined in the paper and that those trying to measure SEA parameters will be able to design better and more accurate measurement techniques given a better understanding of the theory.

In several past papers an effort has been made to verify the validity of SEA concepts (Hodges 1986; Langley 1990). These are important studies. However, in this paper I emphasize more the engineering development of SEA models using theoretically computed SEA parameters based on mobility and impedance functions.

The paper is divided into two major sections. The first section deals with the general process of calculating the SEA parameters using mobility functions. In this section, I assume that the power input to a system or the power transmitted between systems can be represented by the product of single force and velocity amplitudes. This would be the case for a point excitation in a single direction or a point connection between systems at which all but one of the degrees of freedom are constrained. In the second major section of the paper the formulations for the SEA parameters are expanded to deal with excitations and connections that include multiple degrees of freedom and may be extended over a line or area. The emphasis is on a practical approach toward computing coupling loss factors.

2. SEA parameters

In SEA the equations of motion describing a dynamic system are cast in terms of power and energy variables (Lyon 1975). Power balance equations are developed by requiring the overall time-average power input to a system to be equal to the sum of the time-average power dissipated within the system due to damping and the net time-average power transmitted to other systems. The power balance equations are appealing because of their simplicity. The difficulties arise in calculating the parameters that govern the power input and the power transmitted. In the following sections these parameters will be formulated using a general mobility formulation.

(*a*) *Power input*

The power inputs to the SEA subsystems provide the basic forcing functions for the power balance equations. In some cases the power input to a system can be measured or determined empirically. However, because of the difficulty in obtaining a direct measurement of power, a calculation of power input using mobility functions – and their inverse, impedance functions – is often required. The mobility functions can be predicted analytically and in many cases measured mobility functions can be used to improve the confidence in the SEA model. It may also be possible to compute the power input to a dynamic system using a finite element model. The finite element prediction of the power input can then be combined with an SEA model of the structure and acoustic spaces to provide a composite model that combines the advantages of the finite element and the statistical energy analysis methods.

In the following section the time-average power input to mechanical and acoustic systems will be formulated for general point, line, and area excitation sources. First, however, we introduce the basic concepts of mobility analysis. For this introduction we assume that the force and velocity of the system can be represented by single variables, F and V.

The mobility function for a mechanical system is the ratio of the complex amplitude of the response velocity to the complex amplitude of the force acting on the system, where harmonic $e^{+j\omega t}$ time-dependence is assumed. In general, the mobility function depends on frequency. The real part of the mobility is the conductance, which is always positive. The imaginary part of the mobility is the susceptance, which can be negative or positive. A change in the sign convention for the complex time dependence, e.g. from $\exp^{+j\omega t}$ to $\exp^{-j\omega t}$, changes the sign of the imaginary part of the mobility but not the real part.

The power input to a system is the product of force and velocity. The time-average power input can be expressed in terms of the complex amplitudes of the force and velocity as

$$W^{\text{in}} = \tfrac{1}{2}\,\text{Re}(F^*V), \tag{1}$$

where W^{in} is the time-average power input, Re signifies the 'real part of', V is the velocity amplitude, and F^* is the complex conjugate of the force amplitude. If the amplitude of the force acting on the structure is known, the power input can be expressed in terms of the magnitude-squared of the force amplitude and the real part of the mechanical mobility; the conductance, G. The mean-square force acting on the structure, $\langle F^2 \rangle$, is equal to one-half the magnitude-squared of the force amplitude so that the power input can be written in terms of the conductance as

$$W^{\text{in}} = \langle F^2 \rangle G. \tag{2}$$

The formulation of the power input in terms of the forces acting on the system and the conductance is generally useful when the forces are known or can be measured. In other cases the excitation force is not known. However, the excitation can be defined by an imposed velocity. In this case the power input can be expressed in terms of the mean-square velocity, $\langle V^2 \rangle$, and the real part of the mechanical impedance, which is generally referred to as the resistance, R,

$$W^{\text{in}} = \langle V^2 \rangle R. \tag{3}$$

In the most general case, an interaction exists between the excitation and the structure so that neither the excitation force nor an imposed velocity can be directly specified. In this case an equivalent mechanical 'circuit' can be set up to model the excitation (Shearer 1971). The concepts of a Thevinen-equivalent and a Norton-equivalent circuit are borrowed from electrical engineering. The Thevinen equivalent circuit can be used to represent a general source of excitation in terms of a source impedance and its 'blocked-force'; the force applied to the structure if it were rigidly constrained from moving. The Norton equivalent circuit can be used to represent the same excitation source in terms of a source impedance and its 'free-velocity'; the velocity of the source when it is detached from the structure.

By using the Thevinen or blocked-force representation of the excitation, the power input to the structure can be written in terms of the mean-square blocked-force as

$$W^{\text{in}} = \langle F_{\text{b}}^2 \rangle R / |Z_{\text{s}} + Z|^2, \tag{4a}$$

where $\langle F_{\text{b}}^2 \rangle$ is the mean-square blocked force, Z_{s} is the source impedance of the excitation, and Z is the impedance of the excited structure. The power input from the same excitation source can also be written using the Norton or free-velocity representation in terms of the mean-square free-velocity as,

$$W^{\text{in}} = \langle V_{\text{f}}^2 \rangle G / |Y_{\text{s}} + Y|^2, \tag{4b}$$

where $\langle V_{\text{f}}^2 \rangle$ is the mean-square free velocity and Y_{s} is the source mobility (inverse of impedance). As the power input determined from (4*a*, *b*) must be equal we find that the mean-square blocked force and free velocity are related by the magnitude squared of the source impedance,

$$\langle F_{\text{b}}^2 \rangle / \langle V_{\text{f}}^2 \rangle = |Z_{\text{s}}|^2. \tag{4c}$$

This relation can be used as a means to determine the source impedance from measured data or to relate the blocked force and the free velocity in formulations of power input.

The general mobility formulation can be extended to the case where an excitation is applied to the junction of several systems. For this case a junction impedance is defined. Because the systems connected at a structural junction share a common velocity at the junction the junction impedance is the sum of the source impedance plus the impedances of all systems connected to the junction,

$$Z_{\text{jnc}} = Z_{\text{s}} + \sum_i Z_i, \tag{5a}$$

where the summation is over all systems connected at the junction. The power input to one of the connected subsystems can be written in terms of the mean-square blocked force and the junction impedance as

$$W_r^{in} = \langle F_b^2 \rangle R_r / |Z_{jnc}|^2, \qquad (5b)$$

where W_r^{in} is the power input to the receiving system r and R_r is the resistance of system r. The formulation in terms of free velocity and system mobilities is more complex since the mobility of a structural junction is not the sum of the mobilities of the connected systems.

(b) *Random excitation*

When the excitation force is a random function of time, the power input in a band of frequencies can be found by integrating the product of the spectral density of the force and the conductance over the band, $\Delta\omega$,

$$W^{in} = \int_{\Delta\omega} d\omega \, S_F(\omega) \, G(\omega), \qquad (6a)$$

where ω is the radian frequency, W^{in} is the power input in the frequency band, $\Delta\omega$, and $S_F(\omega)$ is the power spectral density of the excitation force. For many cases of practical interest the spectral density of the force can be assumed to be fairly constant over the frequency band $\Delta\omega$. Then the power input can be rewritten in terms of the average conductance and the mean-square force within the band,

$$W^{in} = \langle F^2 \rangle \langle G \rangle_{\Delta\omega}, \qquad (6b)$$

where $\langle \, \rangle_{\Delta\omega}$ signifies an average over the frequency band $\Delta\omega$. The process of averaging over the frequency bandwidth can be easily extended to a velocity source by using a frequency-averaged mechanical resistance in (3).

The use of a frequency-average impedance or mobility function to evaluate the power input for the general excitation represented by a blocked force or a free velocity and a source impedance is not immediately clear. As in the earlier derivation for the force and velocity sources, we should average the input power given by (4) or (5) over a band of frequencies. Although formally correct this averaging process brings about no immediate simplification. It is common in most SEA formulations to use average impedance or average mobilities to evaluate the power input using (4) or (6),

$$W^{in} = \langle F_{b1}^2 \rangle \frac{\langle R \rangle}{|Z_s + \langle Z \rangle|^2} \quad \text{or} \quad W^{in} = \langle V_{fr}^2 \rangle \frac{\langle G \rangle}{|Y_s + \langle Y \rangle|^2}, \qquad (7)$$

where $\langle \, \rangle$ signifies an average over the frequency band. The validity of using an average impedance or mobility to determine the power input is a key question in assessing the accuracy of many SEA parameter formulations. As we will see in a later section the same question arises when computing the coupling loss factors, because an impedance or mobility formulation of the power transmission is generally used.

At very high frequencies, where the average spacing between resonance frequencies is small compared with the response bandwidth of the individual resonances, the mechanical impedance functions will be fairly smooth functions of frequency. In this case the two results given above will be the same, so that using the average impedance or mobility to evaluate the power input is valid.

The use of an average impedance or mobility is also valid if the excitation source impedance is either much greater than or much smaller than the mechanical impedance of the structure. In such cases the source can be modelled either as a velocity or force source and the power input can be evaluated using the average resistance or conductance.

At lower frequencies and for very lightly damped structures, the impedance functions can vary significantly over a band of frequency. At these frequencies the use of the average impedance or mobility to evaluate the power input may result in a biased estimate. Further work in this area is warranted.

(c) *Statistical estimates of the mobility functions*

The use of a frequency-average conductance for a random excitation is a common example where a statistical estimate of a mobility function is used to determine the power input. However, a broader application of the average conductance can be made which is applicable both to random and deterministic single frequency excitations.

In SEA, the structure is described statistically so that the resonance frequencies and mode shapes become random variables. In this case we can define an ensemble of structures and use ensemble averages and other statistical measures to define the structural parameters. Thus, the power input for a force source averaged over the ensemble can be written in terms of an ensemble average conductance,

$$\langle W^{\text{in}} \rangle_{\text{ens}} = \langle F^2 \rangle \langle G(\omega) \rangle_{\text{ens}}, \tag{8}$$

where $\langle G(\omega) \rangle_{\text{ens}}$ is the ensemble-average conductance at the single frequency, ω.

The concept of an ensemble of structures may be difficult to accept for many engineers. It can imply poor manufacturing tolerances and poor quality so that the actual dimensions of the structure vary significantly for the different members of the ensemble. This need not be the case, particularly at frequencies above the first few system resonance frequencies. An ensemble can be formed from identical structures with small changes in operating conditions, temperature, and other variables that cause a seemingly random variation of resonance frequencies and mode shapes. Also during the preliminary design of a product, an ensemble of systems can be formed with variations to account for design uncertainties. As the design matures these uncertainties can be removed and the prediction uncertainties can be removed.

The definition of an ensemble of structures can also be applied when the excitation force is random. In this case the assumption that the force spectral density can be constant over the frequency band need not be made and the ensemble-average conductance can be used to relate the spectral densities of the power input and the force on a narrow-band basis,

$$\langle S_W(\omega) \rangle_{\text{ens}} = S_{\text{F}}(\omega) \langle G(\omega) \rangle_{\text{ens}}, \tag{9}$$

where $S_W(\omega)$ is the spectral density of the power input; formally, the real part of the cross-spectrum of the force and velocity.

The frequency or ensemble average mobility of a system can be formulated in terms of the modes of the system. Following a classical modal analysis the response of the system can be expressed as a sum of the responses of the individual modes,

$$V(x) = \sum_{n=1}^{\infty} V_n \psi_n(x), \tag{10}$$

where V_n is the complex amplitude of the response of the nth mode and $\psi_n(x)$ is the mode shape of the nth mode. The response amplitude, V_n, can be found as the product of the force acting on the mode and the mobility for a single degree-of-freedom system. For a point force acting on the system at point x the modal force amplitude is the product of the applied force amplitude and the value of the mode shape at the application point of the force. The mobility function for a single mode of the system represented by a mass/spring oscillator can be written in terms of the mass, the resonance frequency, and the damping loss factor,

$$Y_n(\omega) = \frac{1}{M_n} \frac{\omega^2 \omega_n \eta_n + j\omega(\omega_n^2 - \omega^2)}{(\omega_n^2 - \omega^2)^2 + (\eta_n \omega_n \omega)^2}, \tag{11a}$$

where M_n is the mass, ω_n is the resonance frequency of the system, η_n is the damping loss factor (viscous damping has been assumed although inclusion of both viscous and solid type damping loss factors is possible). The point mobility of the system can be found as a summation of the response velocities at point x from each of the modes,

$$Y(x, \omega) = \sum_n \psi_n^2(x)\, Y_n(\omega), \tag{11b}$$

where $Y(x, \omega)$ is the point mobility of the system at point x and frequency ω and $Y_n(\omega)$ is the mobility function for the nth mode.

An average mobility over frequency or over an ensemble of systems can be found as the average of the individual terms of the summation,

$$\langle Y(x, \omega)\rangle = \sum_n \langle \psi_n^2(x)\, Y_n(\omega)\rangle, \tag{12a}$$

where $\langle\,\rangle$ represents an average. The average of the product can usually be expressed as the product of the averages. If a frequency average is being formed the mode shapes can be assumed to be frequency independent. If an ensemble average is being formed the mode shapes and the mode mobility function can be assumed to be statistically independent. In either case we can write the average mobility as

$$\langle Y(x, \omega)\rangle = \sum_n \langle \psi_n^2(x)\rangle \langle Y_n(\omega)\rangle. \tag{12b}$$

To evaluate the average we consider first the mobility functions for the individual modes, $Y_n(\omega)$.

For a lightly damped mode, the real part of the mobility shows a large peak at the resonance frequency. The imaginary part shows both a positive and negative peak near the resonance frequency. In evaluating the average mobility we consider first the conductance, the real part of Y. If we integrate the conductance of a single mode over a band of frequencies to determine the frequency-average, we can consider two distinct cases. When the resonance frequency is within the band, the integrated conductance will be large. In this case it is possible to extend the limits of integration to zero and infinity without significantly adding to the value of the integral. The resulting integral is simply $\pi/2M$. Thus, when the frequency band encompasses the resonance frequency of the mode, the frequency-average conductance for the mode is

$$\langle G_n\rangle_{\Delta\omega} = \frac{1}{\Delta\omega} \frac{\pi}{2M_n}, \tag{13}$$

where $\Delta\omega$ is the averaging bandwidth. We find that the average conductance does not depend on resonance frequency except to the extent that we require the frequency to be in the band. This leads to the idea that we do not need to know the exact resonance frequencies of a system, but only the number of modes with resonances in the band and the system mass. In addition, the average conductance does not depend on damping except to the extent that we require the damping to be small (damping loss factors less than 0.3 are sufficiently small that (13) gives a good estimate of the average conductance in the band). When the resonance frequency is outside the averaging band, the average conductance of the mode will be small and its contribution to the summation can generally be ignored.

The imaginary part of the mobility shows both a large positive and a large negative peak for the lightly damped system. At the resonance frequency the imaginary part of the mobility is zero. If we integrate over a frequency band containing the resonance frequency so that both the positive and negative peaks are included, the contribution to the two peaks tends to cancel so that the frequency-average of the imaginary part of the mobility tends to zero.

A frequency average mobility for the system is found by integrating the conductance of each mode over the bandwidth $\Delta\omega$ and summing over all modes. If we limit our attention to the conductance of lightly damped systems, we need consider only those modes with resonance frequencies in the band. For those modes we can extend the limits of integration to zero and infinity in evaluating the integral. The frequency-average point conductance can then be written as a simple summation over modes with resonance frequencies in the band

$$\langle G(x)\rangle_{\Delta\omega} = \frac{1}{\Delta\omega}\frac{\pi}{2}\sum_{\substack{\text{modes}\\ \text{in}\,\Delta\omega}} \frac{\psi_n^2(x)}{M_n}, \tag{14}$$

where $\Delta\omega$ is the averaging bandwidth and the summation is over all modes with resonance frequencies in the band.

Further simplification of the point conductance is possible, although not always justified. Simplification of the power input can result from averaging the conductance over an ensemble of systems. If the ensemble is defined so that the point at which the excitation is applied is a random variable the value of the mode shape squared can be replaced by an average value. For a homogeneous system this average value is simply the ratio of the modal mass divided by the physical mass of the system. The summation is then simply the number of modes with resonance frequencies in the band, $\Delta\omega$,

$$\langle G(x)\rangle_{\text{ens}} = \pi N/2M\,\Delta\omega, \tag{15}$$

where N is the mode count; the number of modes with resonance frequencies in the band. The ratio of the mode count to the bandwidth is the modal density for the system. Thus, we obtain the commonly used expression for the average conductance (Cremer 1988),

$$\langle G(x)\rangle_{\text{ens}} = \pi n(\omega)/2M, \tag{16}$$

where $n(\omega)$ is the modal density-average number of modes per unit radian frequency and M is the physical mass of the system.

The idea of a ensemble of systems with a randomly varying point of excitation is often hard to justify. However, we can also define an ensemble of systems with randomly varying boundary conditions. Because the mode shapes are strongly dependent on boundary conditions, we can achieve the same result for the ensemble average of the mode shape squared.

The relation between the average conductance and the modal density is a key relation in SEA parameter derivatives. However, the assumptions used to derive the simple result in (16) from the more general result in (14) limit the usefulness of the simple result to homogeneous subsystems. Because all dynamic systems are inherently non-homogeneous, it is best to start with the general result in (14) and then determine if further simplification is possible. For example, if the SEA analysis is restricted to high frequencies and large frequency-averaging bandwidths, the excited SEA subsystems can be made sufficiently small that they can be modelled by homogeneous subsystems and (16) can be used. On the other hand, at low frequencies, or for narrow frequency bandwidths, small subsystems may have no modes. In this case it is necessary to increase the size of the excited subsystem and the more general result of (16) may be required.

(*d*) *Coupling loss factor*

In SEA the coupling loss factor relates the power transmitted between two connected subsystems to their energies

$$W^{\text{trans}}_{\text{s;r}} = \omega\eta_{\text{s;r}}E_{\text{s}} - \omega\eta_{\text{r;s}}E_{\text{r}}, \tag{17}$$

where $W^{\text{trans}}_{\text{s;r}}$ is the time-average power transmitted from subsystem s to subsystem r, $\eta_{\text{s;r}}$ is the coupling loss factor between subsystem s and subsystem r, and E_{s} is the total energy of subsystem s. The coupling loss factors are not reciprocal so that $\eta_{\text{s;r}} \neq \eta_{\text{r;s}}$.

To formulate the coupling loss factor using mobility functions we use the power input relations derived in §2*b* together with a relation between the energy of the system and the blocked force or free velocity. If the two systems s and r are disconnected at the junction the resulting velocities of these systems can be considered to be the free velocities in formulating the power transferred between the two systems. In SEA the free velocities of system s and system r are assumed to be uncorrelated. With this assumption the net power transmitted between the systems can now be expressed using the power input formulation in terms of free velocities,

$$W^{\text{trans}}_{\text{s;r}} = \langle V^2_{\text{fs}}\rangle \frac{G_{\text{r}}}{|Y_{\text{s}}+Y_{\text{r}}|^2} - \langle V^2_{\text{fr}}\rangle \frac{G_{\text{s}}}{|Y_{\text{s}}+Y_{\text{r}}|^2}, \tag{18a}$$

where $\langle V^2_{\text{fs}}\rangle$ is the mean-square free velocity of system s and $\langle V^2_{\text{fr}}\rangle^2$ is the mean-square free velocity of system r. Similarly, the power transmitted between the two systems can be written in terms of the mean-square blocked forces,

$$W^{\text{trans}}_{\text{s;r}} = \langle F^2_{\text{bs}}\rangle \frac{R_{\text{r}}}{|Z_{\text{s}}+Z_{\text{r}}|^2} - \langle F^2_{\text{br}}\rangle \frac{R_{\text{s}}}{|Z_{\text{s}}+Z_{\text{r}}|^2}, \tag{18b}$$

where $\langle F^2_{\text{bs}}\rangle$ is the mean-square blocked force of system s and $\langle F^2_{\text{br}}\rangle$ is the mean-square blocked force of system r.

To determine the coupling loss factors it is necessary to simplify (18) by averaging over frequency or over an ensemble of systems. As discussed in §2*b*, it is common in most SEA formulations to use average mobility or impedance functions in (18) to evaluate the transmitted power.

The frequency-average or ensemble-average mean-square free velocities of the two systems are assumed to be proportional to the mean-square kinetic energies of the systems. Because the mean-square kinetic and potential energies of a system are equal for resonant vibration, we can also assume the mean-square free velocities to be proportional to the total energies of the systems,

$$\langle V_{\mathrm{fs}}^2 \rangle = E_{\mathrm{s}}/M_{\mathrm{s}} \quad \text{and} \quad \langle V_{\mathrm{fr}}^2 \rangle = E_{\mathrm{r}}/M_{\mathrm{r}}, \tag{19}$$

where M_{s} and M_{r} are the masses of the two systems.

The process of disconnecting the two systems changes the response of the systems and the distribution of energy. We assume, however, that the relation between the mean-square free velocities and the system energies continues to be valid. With these assumptions the coupling loss factor can be written as

$$\eta_{\mathrm{s;r}} = \frac{1}{\omega M_{\mathrm{s}}} \frac{\langle G_{\mathrm{r}} \rangle}{|\langle Y_{\mathrm{s}} \rangle + \langle Y_{\mathrm{r}} \rangle|^2} \quad \text{and} \quad \eta_{\mathrm{r;s}} = \frac{1}{\omega M_{\mathrm{r}}} \frac{\langle G_{\mathrm{s}} \rangle}{|\langle Y_{\mathrm{s}} \rangle + \langle Y_{\mathrm{r}} \rangle|^2}, \tag{20}$$

where $\langle \rangle$ indicates that an average mobility function is being used. A symmetric coupling factor can be defined as the product,

$$\phi_{\mathrm{s;r}} = \omega \eta_{\mathrm{s;r}} n(\omega)_{\mathrm{s}}, \tag{21}$$

where $\phi_{\mathrm{s;r}}$ is the coupling factor and $n(\omega)$ is the modal density as a function of radian frequency. The coupling factor can be expressed in terms of the average mobility functions as

$$\phi_{\mathrm{s;r}} = \frac{n(\omega)_{\mathrm{s}}}{M_{\mathrm{s}}} \frac{\langle G_{\mathrm{r}} \rangle}{|\langle Y_{\mathrm{s}} \rangle + \langle Y_{\mathrm{r}} \rangle|^2}. \tag{22}$$

The ratio of the modal density to the system mass is proportional to the average conductance of the system. Thus, the coupling factor can be written as

$$\phi_{\mathrm{s;r}} = \frac{1}{2\pi} \frac{4\langle G_{\mathrm{s}} \rangle \langle G_{\mathrm{r}} \rangle}{|\langle Y_{\mathrm{s}} \rangle + \langle Y_{\mathrm{r}} \rangle|^2}. \tag{23a}$$

A similar derivation using the blocked force instead of the free velocity results in an impedance function formulation for the coupling factor

$$\phi_{\mathrm{s;r}} = \frac{1}{2\pi} \frac{4\langle R_{\mathrm{s}} \rangle \langle R_{\mathrm{r}} \rangle}{|\langle Z_{\mathrm{jnc}} \rangle|^2}, \tag{23b}$$

where $\langle Z_{\mathrm{jnc}} \rangle$ is the frequency-average or ensemble-average junction impedance; the sum of the average impedance of all systems connected to the junction.

The coupling factor given by (23*a*, *b*) can be used for one-, two- and three-dimensional systems coupled at a point with a single junction degree of freedom. For the special case of two one-dimensional systems, the mobility ratio in (23*a*) can be replaced by a power transmission coefficient; the ratio of the power transmitted to the receiving system to the power incident from the source system. Although coupling loss factors are often derived from power transmission coefficients, the mobility or impedance formulations given by (23*a*, *b*) are more general.

3. Extended interactions

(*a*) *Point source*

The excitation can be modelled by a point source if its extent is small compared with the wavelengths of vibration in the excited system. The most common example of a point source is a mechanical shaker used to excite a structure at a point.

For the general case of a point excitation of a structural system we must consider all six degrees of freedom; three translational or force degrees of freedom and three rotational or moment degrees of freedom. Because of the orthogonality of the force and moment components in the three axes, the power input can be written as

$$W_{\text{in}} = \tfrac{1}{2}\,\text{Re}\sum_i\sum_j F_i^*\,Y_{i;j}\,F_j, \tag{24a}$$

where the subscripts i and j range from 1 to 6 for the six degrees of freedom, F_i is a force or moment amplitude, $Y_{i;j}$ is the i,j term of the mobility matrix, and F_i^* is the complex conjugate of the force amplitude F_i. The terms of the mobility matrix are the velocity, V_i, divided by the force amplitude, F_j, where V_i and F_j refer to one of the translational and/or rotational degrees of freedom.

From the principle of reciprocity the mobility matrix is symmetric. With this condition the power input can be written in terms of the conductance matrix,

$$W_{\text{in}} = \tfrac{1}{2}\sum_i\sum_j F_i^*\,G_{i;j}\,F_j, \tag{24b}$$

where the imaginary terms cancel.

The cross-terms needed to evaluate the power input can be neglected in an ensemble-average, because both the cross-terms of the excitation and the cross-terms of the conductance matrix can be assumed to average to zero or to a sufficiently small value that they can be neglected. Although some error may be incurred it is generally acceptable in an engineering application to neglect the contribution of the cross-terms so that the power input for the general point force excitation can be written as a simple sum over the six degrees-of-freedom,

$$\langle W_{\text{in}}\rangle = \sum_i \langle F_i^2\rangle\,\langle G_i\rangle, \tag{25}$$

where $\langle F_i^2\rangle$ is the mean-square force (or moment) for the ith degree-of-freedom and $\langle G_i\rangle$ is the average of the ith diagonal term of the conductance matrix.

(*b*) *Line source*

The formulation of power input can be extended to a force excitation distributed over a line by expanding the force into a series of orthogonal functions, usually a complex Fourier series,

$$F(x) = \sum_i F(k_i)\,\mathrm{e}^{-jk_i x}, \tag{26}$$

where $F(k_i)$ is the Fourier amplitude for the component of the force at the wavenumber, k_i. The velocity is also expanded into a Fourier series, so that the intensity at point x (input power per unit length) can be determined from a double summation over Fourier amplitudes,

$$I_{\text{in}}(x) = \tfrac{1}{2}\,\text{Re}\sum_i\sum_j F(k_i)^*\,V(k_j)\,\mathrm{e}^{j(k_i-k_j)\,x}, \tag{27}$$

where $I_{\text{in}}(x)$ is the power intensity at point x. The input power is found by integrating over the length of the excitation. Because of the orthogonality of the Fourier functions, the cross-terms drop out so that the power input can be written in terms of a single summation over the Fourier amplitudes,

$$W_{\text{in}} = \tfrac{1}{2}L\,\text{Re}\sum_i F(k_i)^*\,V(k_i), \tag{28}$$

where L is the length of the distributed force excitation. The Fourier amplitude of the response velocity, $V(k_i)$, can be expressed as the matrix product of a mobility matrix and the Fourier amplitudes of the applied force. Because of reciprocity the mobility matrix relating the complex Fourier amplitudes of the response velocity and the excitation force must be symmetric. Then the complex mobility matrix can be replaced by the conductance matrix, which is real. The power input can now be written as

$$W_{\text{in}} = \tfrac{1}{2}L \sum_i \sum_j F(k_i)^* \, G(k_i, k_j) \, F(k_j), \tag{29}$$

where the imaginary parts of the matrix terms cancel. A term of the conductance matrix, $G(k_i, k_j)$, is the ratio of the complex Fourier amplitude of the response velocity at wavenumber k_i to the complex Fourier amplitude of the excitation force at wavenumber k_j.

The terms of the conductance matrix can be formulated in terms of the modes of the system as,

$$\langle G(k_i, k_j) \rangle_{\Delta\omega} = \frac{1}{\Delta\omega} \frac{\pi}{2} \sum_n \frac{\tilde{\psi}_n(k_i) \, \tilde{\psi}_n(k_j)^*}{M_n}, \tag{30a}$$

where the summation is over all modes with resonance frequencies in the band $\Delta\omega$ and the Fourier transform of the mode shape is defined as

$$\tilde{\psi}_n(k_i) = \frac{1}{L} \int_0^L \mathrm{d}x \, \psi_n(x) \, \mathrm{e}^{jk_i x}. \tag{30b}$$

In forming an ensemble average of the power input the cross terms of the conductance matrix can be neglected, because they can be assumed to average to zero. Thus the average power input from the line excitation can be written as

$$\langle W_{\text{in}} \rangle = L \sum_i \langle F(k_i)^2 \rangle \langle G(k_i) \rangle, \tag{29a}$$

where $\langle F(k_i)^2 \rangle$ is the mean-square value of the Fourier component at k_i and $\langle G(k_i) \rangle$ is the average of a diagonal term of the conductance matrix.

(c) *Area source*

The power input from a pressure field associated with an acoustic source or a turbulent boundary layer is determined by a similar approach to that used of the line-distributed excitation. The area-distributed pressure field is expanded into Fourier amplitudes in two dimensions,

$$F(x, y) = \sum_{ix} \sum_{iy} F(k_{ix}, k_{iy}) \, \mathrm{e}^{-jk_{ix}x} \, \mathrm{e}^{-jk_{iy}y}, \tag{31}$$

where $F(k_{ix}, k_{iy})$ is the Fourier amplitude for the component of the force at the wavenumber defined by the components, k_{ix} and k_{iy}. The velocity is also expanded into its Fourier amplitudes and following the approach used for the line-distributed excitation the power input is found to be,

$$W_{\text{in}} = \tfrac{1}{2}A \sum_{ix} \sum_{iy} \sum_{jx} \sum_{jy} F(k_{ix}, k_{iy})^* \, G(k_{ix}, k_{iy}; k_{jx}, k_{jy}) \, F(k_{jx}, k_{jy}), \tag{32}$$

where the imaginary parts of the matrix terms cancel. The conductance matrix,

$G(k_{ix}, k_{iy}; k_{jx}, k_{jy})$ is the real part of the ratio of the complex Fourier amplitude of the response velocity at the wavenumber defined by the components k_{ix} and k_{iy} to the complex amplitude of the excitation force at the wavenumber defined by the components k_{jx} and k_{jy}. As for the case of a line excitation the cross terms can be neglected in an ensemble average of the power input so that

$$\langle W_{\text{in}} \rangle = A \sum_{ix} \sum_{iy} \langle F(k_{ix}, k_{iy})^2 \rangle \langle G(k_{ix}, k_{iy}) \rangle, \tag{33}$$

where $\langle F(k_{ix}, k_{iy})^2 \rangle$ is the mean-square value of the Fourier component at k_{ix} and k_{iy} and $\langle G(k_{ix}, k_{iy}) \rangle$ is the average of a diagonal term of the conductance matrix.

4. Conclusions

The use of power and energy variables to describe the dynamic response of structural-acoustic systems is a key concept of SEA. In this paper the power input to a subsystem and the power transmitted between connected systems have been formulated using mobility functions. This in itself does not represent an advance to SEA theory. In fact, nearly all published derivations of SEA parameters are based, at least implicitly, on a mobility formulation. Because of the complexity of the general mobility formulation for most structural-acoustic problems few exact solutions for SEA parameters exist. The difficulty comes in identifying the approximations that are made and their appropriateness to the problems being considered. In this paper we have shown that a generalized and simplified formulation of the power input and coupling loss factor can be achieved by averaging either over frequency or over an ensemble of systems. The averaging process results in two general simplifications. First, it is usually possible to ignore the cross-terms in the mobility formulations. Second, it is possible to relate the conductance to the modal density and the mass of the system. A formal proof of the validity of these simplifications is not always possible. However, by making these simplifications in computing the SEA parameters complex dynamic problems that are otherwise intractable can be analysed using SEA.

References

Cremer, L. & Heckl, M. 1988 Point excitation of finite systems. In *Structure-borne sound*, pp. 330–332. Berlin: Springer-Verlag.

Hodges, C. H. & Woodhouse, J. 1986 Theories of noise and vibration transmission in complex structures. *Rep. Prog. Phys.* **49**, 107–170.

Langley, R. S. 1990 A derivation of the coupling loss factors used in statistical energy analysis. *J. Sound Vib.* **141**, pp. 207–219. Academic Press.

Lyon, R. H. 1975 *Statistical energy analysis of vibrating systems*. Cambridge, Massachusetts: MIT Press.

Manning, J. E. 1990 Calculation of statistical energy analysis parameters using finite element and boundary element models. In *Proc. Int. Congr. on Recent Developments in Air and Structure-Borne Sound and Vibration*. Auburn University.

Rockwood, W. B. *et al.* 1987 Statistical energy analysis applied to structure borne noise in marine structures. In *Statistical energy analysis* (ed. K. H. Hsu, D. J. Nefske & A. Akay), NCA-Vol. 3, pp. 73–79. New York: The American Society of Mechanical Engineers.

Shearer, J. L. *et al.* 1971 Equivalent networks. In *Introduction to system dynamics*, pp. 348–356. Reading, Massachusetts: Addison-Wesley.

Wave intensity analysis of high frequency vibrations

By Robin S. Langley[1] and Ahmet N. Bercin[2]

[1] *Department of Aeronautics and Astronautics, University of Southampton, Southampton SO9 5NH, U.K.*

[2] *College of Aeronautics, Cranfield Institute of Technology, Cranfield, Bedford MD43 0AL, U.K.*

In the statistical energy analysis (SEA) approach to high frequency dynamics it is assumed that the vibrational wavefield in each component of an engineering structure is diffuse. In some instances the directional filtering effects of structural joints can lead to highly non-diffuse wavefields, and in such cases SEA will yield a very poor estimate of the vibrational response. An alternative approach is presented here in which the directional dependency of the vibrational wavefield in each component is modelled by using a Fourier series. It is shown that, if required, the resulting energy balance equations may be cast in the form of conventional SEA with the addition of 'non-direct' coupling loss factors. The method is applied to the bending and in-plane vibrations of various plate structures and a comparison is made with exact results yielded by the dynamic stiffness method. A significant improvement over conventional SEA is demonstrated.

1. Introduction

The analysis of high-frequency vibration levels in engineering structures causes severe difficulties for standard analysis procedures such as the finite element method (Zienkiewicz 1977). This approach, and all others that are based on the solution of the governing constitutive equations, requires an excessive number of degrees of freedom to capture the short wavelength structure deformation that occurs at high frequencies. A well-known alternative approach is statistical energy analysis (SEA) (Lyon 1975) in which a complex structure is represented as an assembly of subsystems whose vibrational energy levels are calculated from power balance considerations.

A central tenet of SEA is that within a subsystem there is 'equipartition' of vibrational energy among the constituent modes. From a wave, rather than a modal, point of view, this principle is equivalent to assuming that the vibrational wavefield in each structural element is diffuse. Although this assumption may be reasonable in many cases, there are instances where the directional filtering effect of structural junctions can lead to wavefields that are far from diffuse; in such cases SEA yields a very poor estimate of the vibrational energy levels (Blakemore *et al.* 1990; Langley 1992). In this paper an alternative approach is presented in which the directional dependence of the vibrational wave intensity in each structural component is modelled by using a Fourier series; if a single Fourier term is used, then conventional

SEA is recovered. This approach, which is referred to as wave intensity analysis (WIA), was first developed by Langley (1992) who demonstrated that the method can yield much improved estimates for the bending vibrations of plate structures.

It is shown here that the power balance equations that arise in WIA may, if required, be cast in the form of conventional SEA with the addition of 'non-direct' coupling loss factors between subsystems which are not physically connected. This is consistent with the analysis of Langley (1989*b*, 1990) who demonstrated that non-direct coupling loss factors may be of significant effect even for structures which satisfy the conditions which are normally laid down for the successful application of conventional SEA. The method is applied here to a range of plate structures, and a comparison is made with exact calculations based on the dynamic stiffness method. Whereas Langley (1992) considered only bending vibrations, both bending and in-plane vibrations are considered here, and a significant improvement over conventional SEA is demonstrated.

2. Wave intensity analysis

Engineering structures are normally composed of a number of regular structural elements such as beams, plates, and shells, which are either bolted, welded, or bonded together. The differential equation which governs the dynamic response of a typical element may be written in the general form

$$\boldsymbol{L}(\boldsymbol{w})-\rho\,\partial^2\boldsymbol{w}/\partial t^2 = \boldsymbol{F}(\boldsymbol{x},t), \tag{2.1}$$

where $\boldsymbol{w}(\boldsymbol{x},t)$ is the displacement vector at the spatial location $\boldsymbol{x}$, ρ is the mass density, and $\boldsymbol{F}(\boldsymbol{x},t)$ is the applied distributed loading. Further $\boldsymbol{L}$ is a structural differential operator which characterizes the component: in what follows it will be assumed that each component is homogeneous, so that $\boldsymbol{L}$ does not vary with spatial position. Equation (2.1) can be used to determine the various types of elastic wave which can be borne by the component. The concern here is with harmonic plane waves which have the form $\boldsymbol{w} = \boldsymbol{a}\exp(\mathrm{i}\boldsymbol{k}.\boldsymbol{x}-\mathrm{i}\omega t)$, where $\boldsymbol{k}$ is the wavenumber vector, ω is the circular frequency, and $\boldsymbol{a}$ is the wave 'mode'. Valid solutions for $\boldsymbol{k}$ and $\boldsymbol{a}$ at a specified frequency ω may be sought by substituting the assumed wave form into the homogeneous version of equation (2.1). This yields

$$[\boldsymbol{\Lambda}(\boldsymbol{k})-\rho\omega^2\boldsymbol{I}]\,\boldsymbol{a} = \boldsymbol{0}, \tag{2.2}$$

where $\boldsymbol{I}$ is the identity matrix and the detailed form of the matrix $\boldsymbol{\Lambda}$ is determined by the operator $\boldsymbol{L}$. The n-dimensional wavenumber vector $\boldsymbol{k}$ may be expressed in terms of the scalar wavenumber $\mu = |\boldsymbol{k}|$ and a set of $n-1$ direction cosines, $\boldsymbol{\theta}$ say. With this notation equation (2.2) takes the form

$$[\boldsymbol{\Lambda}(\mu,\boldsymbol{\theta})-\rho\omega^2\boldsymbol{I}]\,\boldsymbol{a} = \boldsymbol{0}. \tag{2.3}$$

For specified ω and $\boldsymbol{\theta}$, this is an eigenproblem that may be solved to yield the set of eigenvalues μ_j and eigenvectors $\boldsymbol{a}_j$. The number of distinct real values of μ thus obtained is equal to the number of distinct types of propagating plane wave of heading $\boldsymbol{\theta}$ that can be borne by the element. For an isotropic element equation (2.3) is independent of $\boldsymbol{\theta}$ and the various wavetypes may propagate in all directions. An example in this category is a flat plate element that will display three distinct values of μ, corresponding to out-of-plane bending waves, in-plane shear waves, and in-plane longitudinal waves.

In what follows the high-frequency vibrational response of an engineering structure is described purely in terms of elastic waves. It is assumed that the harmonic response of a typical element may be written in the form

$$\boldsymbol{w}(\boldsymbol{x},t) = \sum_j \int_\theta f_j(\boldsymbol{\theta})\, \boldsymbol{a}_j(\boldsymbol{\theta}) \exp\left[i\boldsymbol{k}(\mu_j,\boldsymbol{\theta})\,.\,\boldsymbol{x} - \mathrm{i}\omega t\right] \mathrm{d}\boldsymbol{\theta}, \tag{2.4}$$

where the sum is over the number of distinct wavetypes displayed by the element and the integral is over the range of possible wave headings; function $f_j(\boldsymbol{\theta})$ represents the amplitude of wavetype j at heading $\boldsymbol{\theta}$. It is further assumed that all phase effects may be neglected, which implies that the various wave components may be considered to be uncorrelated. This assumption has two consequences: first, the response in each component as given by equation (2.4) will be homogeneous with respect to the location $\boldsymbol{x}$, and second, resonance peaks and anti-resonance troughs which are caused by phase effects will not arise. There is evidence to suggest that at high frequencies the response does tend to become homogeneous providing that the element is reverberant, has a high modal overlap factor, and a high number of modes are excited: this aspect has been studied both theoretically and experimentally by Dowell & Kubota (1985, 1986). Further, at high modal overlap the resonant peaks do not differ greatly from the frequency average response level, which implies that phase effects are not significant; at low modal overlap the present approach can be expected to yield a result which corresponds to the average response over a frequency band which contains a number of resonant peaks. If the assumption of uncorrelated wave components is adopted, then equation (2.4) implies that the vibrational energy density within a typical element may be written in the form

$$e(\omega) = \sum_j \int_\theta e_j(\boldsymbol{\theta},\omega)\, \mathrm{d}\boldsymbol{\theta}, \tag{2.5}$$

where $e_j(\boldsymbol{\theta},\omega)$ is the energy density associated with wavetype j at heading $\boldsymbol{\theta}$. The quantities $e_j(\boldsymbol{\theta},\omega)$ are taken to be the basic unknowns in the present analysis method, and a solution is sought by considering power balance.

Although $e_j(\boldsymbol{\theta},\omega)$ has previously been used to represent the energy density of the jth wavetype in a particular component, the notation may be extended so that j is considered to range over all wavetypes in all components of the built up structure. Thus, for example, in the case of a structure consisting of two coupled plates j would range from 1 to 6, covering the three wavetypes which occur in each plate. For each wavetype j the power input through external forcing, P_j^{in} say, together with the power input over the element boundaries P_j^{ci}, must be balanced by the dissipated power P_j^{diss} and the power output over the element boundaries P_j^{co}. Thus

$$P_j^{\mathrm{in}}(\boldsymbol{\theta},\omega) = P_j^{\mathrm{diss}}(\boldsymbol{\theta},\omega) + P_j^{\mathrm{co}}(\boldsymbol{\theta},\omega) - P_j^{\mathrm{ci}}(\boldsymbol{\theta},\omega). \tag{2.6}$$

Now most theoretical models of damping imply that the dissipated power is proportional to the mean stored energy, so that

$$P_j^{\mathrm{diss}}(\boldsymbol{\theta},\omega) = \omega\eta_j A_j e_j(\boldsymbol{\theta},w) = \omega\eta_j E_j(\boldsymbol{\theta},\omega), \tag{2.7}$$

where η_j is the loss factor, A_j is the length, area, or volume of the relevant structural element, and $E_j = A_j e_j$ is the total energy stored in wavetype j at heading $\boldsymbol{\theta}$. The power output by wavetype j at a boundary of the element may be written in the form

$$P_j^{\mathrm{co}}(\boldsymbol{\theta},\omega) = e_j(\boldsymbol{\theta},\omega)\, c_{gj}(\boldsymbol{\theta},\omega) \cos(\delta)\, L, \tag{2.8}$$

where c_{gj} is the group velocity of the wave, δ is the angle between the wave heading and the outward pointing normal to the boundary (assumed to be constant over the boundary), and L is the appropriate dimension (unity, length or area) of the boundary. As the concern is with power output, equation (2.8) relates only to those wave headings for which $0 \leqslant \delta \leqslant \frac{1}{2}\pi$. For clarity, the present analysis is now limited to two-dimensional isotropic components, although in general the method is equally applicable to the full range of structural types. It is also convenient at this stage to introduce the modal density ν_j which is associated with wavetype j for a two-dimensional isotropic element: $\nu_j = \omega A_j / 2\pi c_j c_{gj}$, where c_j is the wave phase velocity and A_j is the area of the element. Equation (2.8) can now be written in the form

$$P_j^{\mathrm{co}}(\theta, \omega) = (\omega L/2\pi)[E_j(\theta, \omega)/\nu_j][\cos(\theta + \tfrac{1}{2}\pi - \psi)/c_j], \tag{2.9}$$

where the angle ψ is used to describe the orientation of the boundary, as shown in figure 1*a*.

It can be noted that in two dimensions the direction cosine vector $\boldsymbol{\theta}$ reduces to the scalar wave heading θ. Equation (2.9) relates to a single output boundary: more generally the result must be summed over all boundaries of the element for which $0 \leqslant \delta \leqslant \frac{1}{2}\pi$.

The final term which appears in equation (2.6) is the power input to heading θ of wavetype j at the element boundaries. This power will be supplied by one or more of the other wavetypes; for example, figure 1*b* illustrates the case in which the power is supplied by heading ϕ of wavetype i. The power input to a band $\mathrm{d}\theta$ of wavetype j may in this case be written in the form

$$P_j^{\mathrm{ci}}(\theta, \omega)\,\mathrm{d}\theta = e_i(\phi, \omega)\, c_{gi}(\omega) \cos(\phi + \tfrac{1}{2}\pi - \psi)\, L \tau_{ij}(\phi + \tfrac{1}{2}\pi - \psi)\,\mathrm{d}\phi, \tag{2.10}$$

where τ_{ij} is the transmission coefficient between the two wavetypes. By making use of Snell's Law and introducing the modal density of wavetype i, equation (2.10) may be rewritten as

$$P_j^{\mathrm{ci}}(\theta, \omega) = (\omega L/2\pi)[E_i(\phi, \omega)/\nu_i][\cos(\theta + \tfrac{1}{2}\pi - \psi)/c_j]\, \tau_{ij}(\phi + \tfrac{1}{2}\pi - \psi). \tag{2.11}$$

This result relates to a single input wavetype and a single input boundary; more generally the total power input may be obtained by summing over the relevant boundaries and wavetypes.

By combining equations (2.6), (2.7), (2.9) and (2.11), the total power balance equation for heading θ of wavetype j may be written in the form

$$P_j^{\mathrm{in}}(\theta, \omega) = \omega \eta_j E_j(\theta, \omega) + (\omega/2\pi c_j)[E_j(\theta, \omega)/\nu_j] \sum_k L_k \cos(\theta + \tfrac{1}{2}\pi - \psi_k)$$

$$- (\omega/2\pi c_j) \sum_m \sum_i [E_i(\phi_{mi}, \omega)/\nu_i] L_m \cos(\theta + \tfrac{1}{2}\pi - \psi_m)\, \tau_{ij}^m(\phi_{mi} + \tfrac{1}{2}\pi - \psi_m), \tag{2.12}$$

where the sum over k represents the output boundaries and the sums over m and i correspond to the input boundaries and the input wavetypes. An equation of this form may be derived for each wavetype j and in principle the set of coupled equations can then be solved to yield the wave energies $E_j(\theta, \omega)$. A convenient approach to the solution of these equations is to expand the angular dependency of the wave energies in the form of a Fourier series so that

$$E_j(\theta, \omega) = \sum_p E_{jp}(\omega) N_p(\theta), \tag{2.13}$$

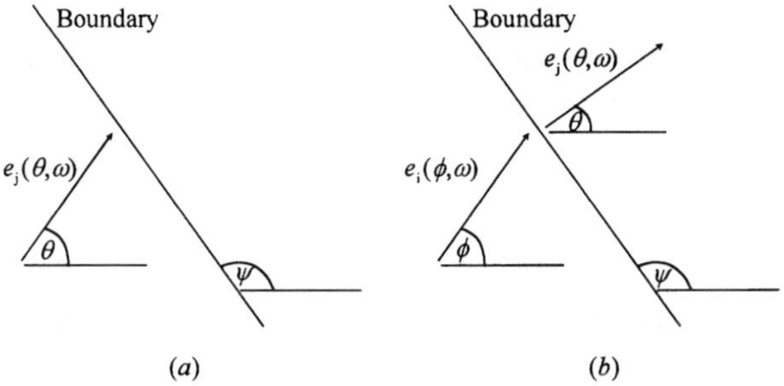

Figure 1. Schematic of wave incidence at a boundary.

where the shape functions $N_p(\theta)$ represent the $\cos(m\theta)$ and $\sin(m\theta)$ Fourier terms. By substituting (2.13) into (2.12) and using the Galerkin procedure, the following set of linear equations may be derived

$$\boldsymbol{C}\hat{\boldsymbol{E}} = \boldsymbol{P}. \tag{2.14}$$

Here the vector $\hat{\boldsymbol{E}}$ contains the coefficients E_{jp}/ν_j and the entries of the matrix $\boldsymbol{C}$ and the $\boldsymbol{P}$ are given by

$$C_{jp,is} = \delta_{ij}\left\{\omega\eta_j\nu_j\int_0^{2\pi} N_p(\theta)N_s(\theta)\,\mathrm{d}\theta + (\omega/2\pi c_j)\sum_k L_k\int_{\theta_k} N_p(\theta)N_s(\theta)\cos(\theta+\tfrac{1}{2}\pi-\psi_k)\,\mathrm{d}\theta\right\}$$

$$-(\omega/2\pi c_j)\sum_m L_m\int_{\theta_m} N_p(\theta)N_s(\phi_{mi})\cos(\theta+\tfrac{1}{2}\pi-\psi_m)\,\tau_{ij}^m(\phi_{mi}+\tfrac{1}{2}\pi-\psi_m)\,\mathrm{d}\theta, \tag{2.15}$$

$$P_{jp} = \int_0^{2\pi} P_j^{\mathrm{in}}(\theta,\omega)N_p(\theta)\,\mathrm{d}\theta. \tag{2.16}$$

As presented above, the matrix $\boldsymbol{C}$ is not symmetric. The Fourier shape functions can be grouped into those for which $N(\theta) = N(\theta+\pi)$ (type A, say) and those for which $N(\theta) = -N(\theta+\pi)$ (type B, say); it can be shown that coupling terms involving two type A or two type B Fourier components are symmetric, while those involving a type A and a type B Fourier component are skew-symmetric. The matrix may readily be made symmetric by multiplying those rows corresponding to type B Fourier components by -1. In what follows it will be assumed that this has been done; the resulting symmetry of the matrix has important implications, as discussed in §3*b*. The relation between the present approach and conventional SEA is discussed in the following section.

3. Relation to statistical energy analysis

(*a*) *Conventional SEA*

The simplest approximation to the angular dependence of the wave energies $E_j(\theta,\omega)$ is to assume that each wave field is diffuse, so that $E_j(\theta,\omega) = E_{j1}(\omega)/2\pi$, which corresponds to the use of a single Fourier term in equation (2.13). If both sides of equation (2.14) are in this case multiplied by 2π then it has been shown by Langley (1992) that a typical coupling term in the matrix $\boldsymbol{C}$ will take the form

$$C_{j1,i1} = -\omega\nu_j\sum_m \eta_{ji}^m, \quad \eta_{ji}^m = L_m c_{gj}\langle\tau_{ji}^m\rangle/(\omega A_j\pi). \tag{3.1, 3.2}$$

Here η_{ij}^m is the coupling loss factor as used in SEA, and $\langle \tau_{ij}^m \rangle$ is the diffuse wave field transmission coefficient (Lyon 1975). Similarly, Langley (1992) has shown that a diagonal term in the matrix $\boldsymbol{C}$ will take the form

$$C_{j1,j1} = \omega \eta_j \nu_j + \omega \nu_j \sum_m \sum_{i \neq j} \eta_{ji}^m. \tag{3.3}$$

Equations (3.1) and (3.3), together with equation (2.14) constitute conventional SEA. It thus follows that the present method reduces to standard SEA if a single Fourier component is used in equation (2.13): in this regard the present approach may be considered to be a natural generalization of SEA.

(*b*) *SEA with non-direct coupling loss factors*

It has been suggested that in some applications SEA should include coupling loss factors between subsystems which are not directly coupled (for example, Blakemore *et al.* 1990). In fact Langley (1990) has demonstrated that in general these coupling loss factors are non-zero, and further there is no sound theoretical reason for assuming that they are negligibly small. However, the calculation of these terms presents severe difficulties and a general methodology has yet to appear in the literature. It is shown in this section that WIA provides just such a methodology: equations (2.14)–(2.16) may be recast in the form of conventional SEA with the addition of non-direct coupling loss factors.

The energy vector $\hat{\boldsymbol{E}}$ which appears in equation (2.14) may be partitioned in the form $\hat{\boldsymbol{E}} = (\hat{\boldsymbol{E}}_1 : \hat{\boldsymbol{E}}_n)$ where $\hat{\boldsymbol{E}}_1$ contains the diffuse energy components: i.e. those components which are related to the first (constant) term in the Fourier expansion of the angular distribution of the wave energy. The partition $\hat{\boldsymbol{E}}_n$ contains the energy terms which are associated with the second and subsequent terms in the Fourier expansion. Two points regarding $\hat{\boldsymbol{E}}_1$ can be noted: first, the terms $\hat{\boldsymbol{E}}_1$ are exactly the energy variables which appear in SEA and second, knowledge of $\hat{\boldsymbol{E}}_1$ alone is sufficient to yield the total energy in an element, as the integral over θ of all but the first Fourier term is zero. The central WIA equation, equation (2.14), may be partitioned in terms of $\hat{\boldsymbol{E}}_1$ and $\hat{\boldsymbol{E}}_n$ to yield

$$\begin{pmatrix} \boldsymbol{C}_{11} & \boldsymbol{C}_{1n} \\ \boldsymbol{C}_{1n}^{\mathrm{T}} & \boldsymbol{C}_{nn} \end{pmatrix} \begin{pmatrix} \hat{\boldsymbol{E}}_1 \\ \hat{\boldsymbol{E}}_n \end{pmatrix} = \begin{pmatrix} \boldsymbol{P}_1 \\ \boldsymbol{P}_n \end{pmatrix}. \tag{3.4}$$

The symmetry of $\boldsymbol{C}$, as noted in §2, has been employed in this equation. Given the definition of $\hat{\boldsymbol{E}}_1$ it follows that $\boldsymbol{C}_{11}$ is precisely the matrix which appears in conventional SEA. Equation (3.4) can be recast in the form

$$(\boldsymbol{C}_{11} - \boldsymbol{C}_{1n} \boldsymbol{C}_{nn}^{-1} \boldsymbol{C}_{1n}^{\mathrm{T}}) \hat{\boldsymbol{E}}_1 = \boldsymbol{P}_1 - \boldsymbol{C}_{1n} \boldsymbol{C}_{nn} \boldsymbol{P}_n. \tag{3.5}$$

The result is essentially SEA with the addition of the triple matrix product which appears on the left-hand side: this symmetric matrix will contain non-direct coupling loss factors. The additional power term which appears on the right-hand side will generally be small or zero, since for most types of loading the input power tends to be fairly diffuse, so that $\boldsymbol{P}_n \approx 0$. Three important conclusions follow from equation (3.5): (i) the non-direct coupling loss factors are independent of the applied loading, (ii) the non-direct coupling loss factors are dependent on the system loss factors, as the diagonals of $\boldsymbol{C}_{nn}$ involve these terms, and (iii) the present approach provides a reasonably straightforward method by which the non-direct coupling loss factors may be calculated.

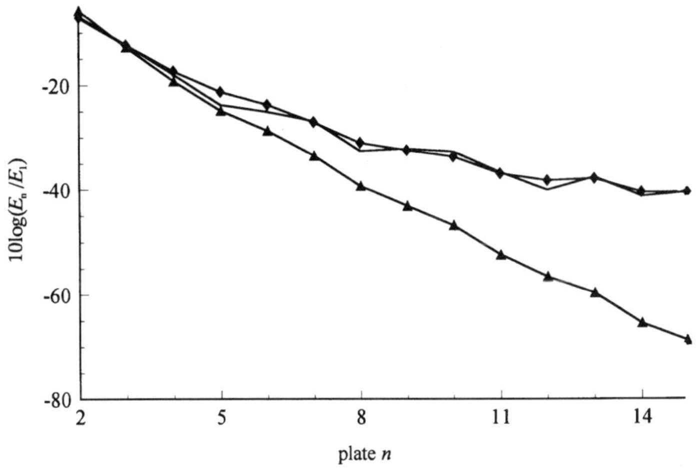

Figure 2. Bending energy distribution in the fifteen plate structure for the one-third octave band with centre frequency 5 kHz. Unmarked curve: exact results. Diamond symbols: WIA results. Triangle symbols: SEA results.

4. Example applications

In this section the foregoing theory is applied to two example structures. The first consists of a chain of 15 plates which are all of width 0.85 m. Each plate is rigidly attached at 90° to its neighbours in the chain so that every second plate is vertical while the remaining plates are horizontal. The structure is taken to be simply supported along the two longitudinal edges and clamped along the two transverse edges. The lengths of the plates (in metres) are 0.9, 1.37, 1.45, 1.1, 0.7, 1.1, 0.9, 0.55, 0.75, 1.1, 0.7, 0.65, 0.95, 0.78 and 1.15. The thicknesses of the plates (in millimetres) are 10, 6.5 8.4, 3.6, 4.5, 6.3, 5.5, 3.0, 4.4, 5.6, 6.0, 4.0, 3.5, 5.2 and 6.5. Each plate is made from steel, which has Young's modulus $E = 2 \times 10^{11}$ N m^{-2}, mass density $\rho =$ 7800 kg m^{-3}, and Poisson ratio $\nu = 0.3$. Exact results for the dynamic response of this structure have been obtained by using the direct dynamic stiffness method. The formulation used represents an extension of the work of Langley (1989*a*) to the case of in-plane vibrations: full details are given by Bercin (1993). The response of the structure to an out-of-plane harmonic point load applied to the first plate has been computed over the frequency range 0.5–20 kHz, and the results obtained have been averaged over nine randomly selected point load locations in the first plate. The average vibrational energy of each plate thus obtained has then been further averaged over sixteen one-third octave bands to yield a frequency response curve which may be compared with SEA and WIA. The modal densities and group velocities that are required by SEA and WIA are standard for plate structures (see, for example, Cremer *et al.* 1988), while the junction wave transmission coefficients have been calculated by using the method of Langley & Heron (1990). Because of symmetry only cosine Fourier terms were used in the WIA technique, and five terms were used for the inner plates while three terms were used for the two outmost plates.

Results for the bending vibrational energy of each plate for the one-third octave band with centre frequency 5 kHz are shown in figure 2. It is clear that the WIA approach yields a very good estimate of the response, whereas SEA tends to overestimate the response for plate 2 and severely underestimate the response for subsequent plates. This is because the wave filtering effect of the structural junctions

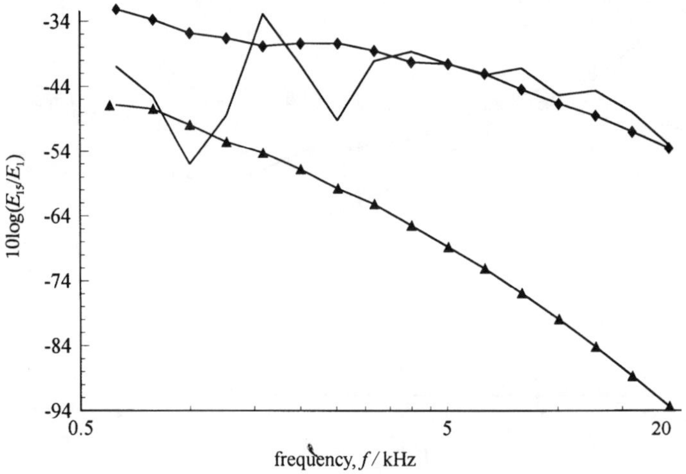

Figure 3. Bending energy in the last plate of the fifteen plate structure. Unmarked curve: exact results. Diamond symbols: WIA results. Triangle symbols: SEA results.

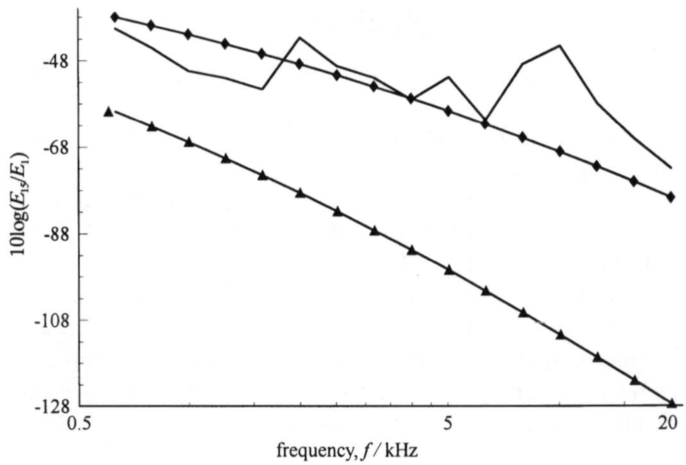

Figure 4. Bending energy in the last plate of the fifteen plate structure with simple support coupling rather than direct connections. Unmarked curve: exact results. Diamond symbols: WIA results. Triangle symbols: SEA results.

is not considered by SEA: waves that are near normal to the various junctions tend to have a high transmission coefficient in comparison with other wave headings, and thus the wavefield becomes less and less diffuse as the vibration travels down the structure. Results for the bending vibrational energy of the final plate over the full frequency range are shown in figure 3, where the under prediction arising from SEA is again apparent. For comparison, results for 'bending only' transmission are shown in figure 4; in this case the structure is taken to be 'flat' with line simple support connections, so that no in-plane waves are generated. The differences between figures 3 and 4 are due solely to the effects of in-plane waves. As might be expected the presence of in-plane waves increases the bending vibrational energy of the structure, as the in-plane dynamics provides an additional 'flanking path' for vibration transmission.

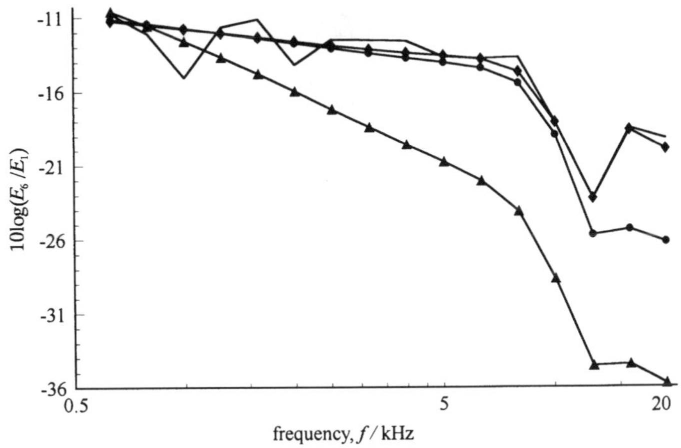

Figure 5. Bending energy in the last plate of the stiffened plate array. Unmarked curve: exact results. Diamond symbols: WIA results for a high number of Fourier terms. Circle symbols: WIA results for a low number of Fourier terms. Triangle symbols: SEA results.

The second example structure consists of a flat row of six plates which are coupled via stiffeners. Each plate has width 0.5 m and thickness 2 mm while the plate lengths (in metres) are 0.3, 0.26, 0.33, 0.28, 0.24 and 0.36. In this case the structure is considered to be made from aluminium which has Young's modulus $E = 7.3 \times 10^{10}$ N m^{-2}, mass density $\rho = 2800$ kg m^{-3}, and Poisson ratio $\nu = 0.33$. The plates are coupled through *symmetrically arranged* rectangular stiffeners of thickness 3 mm and height 200 mm. As in the previous example, the first plate is subjected to an out-of-plane harmonic point load and the dynamic stiffness method has been used to compute the third octave band energy levels. For this example the symmetric arrangement of the stringers prevents the generation of in-plane waves; in this case the aim was to investigate the role of the elastic coupling without the complicating effects of the in-plane dynamics. The vibrational energy in the final plate is shown in figure 5, where it can be seen that SEA again leads to a severe under prediction of the response. Two curves are shown for WIA; one corresponds to the use of five Fourier terms in the inner plate and three Fourier terms in the two outmost plates, while the other corresponds to the use of 41 and 20 Fourier terms respectively. It can be seen that the use of relatively few terms is adequate until around 10 kHz, beyond which the additional terms are needed to capture the behaviour of the exact results. The dip in the response curve at around 15 kHz is related to the behaviour of the wave transmission coefficient of the junction; it should be noted that simple beam theory has been used to model the stiffeners, so that internal dynamic effects have not been considered. The angular dependency of the wavefield in the final plate at 20 kHz is shown in figure 6. Here θ is defined such that $\theta = 0$ represents a wave which is normal to the junction line. The exact curve which is shown in this figure has been deduced from the dynamic stiffness analysis; with this approach the lateral dependency of the response is modelled by a Fourier sine series. As the wavelength of the bending waves at any particular frequency is known, it is possible to compute the heading of the waves whose projected wavelength corresponds to a particular Fourier sine component. A plot of vibrational energy against Fourier sine component may thus be converted to a plot of vibrational energy against wave heading. It is clear from figure 6 that the wavefield is far from diffuse, and good agreement between

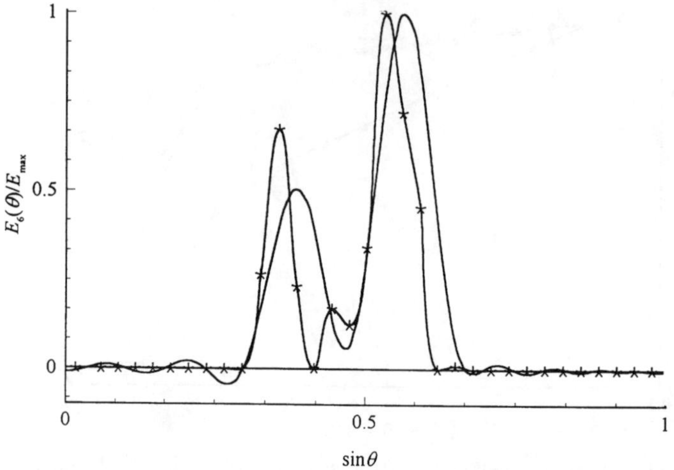

Figure 6. Bending energy variation with wave heading for the final plate of the stiffened plate array. One-third octave band with centre frequency 20 kHz. Unmarked curve: WIA results. Star symbols: exact results.

the exact and WIA results is found. The small region of negative energy that is predicted by WIA may be corrected by the addition of further Fourier terms.

Further examples which cover a wide range of structures have been presented by Bercin (1993).

5. Conclusions

The wave intensity technique which has been presented here is a natural extension of conventional SEA. It has been shown that the method can yield much improved response predictions over SEA with relatively little additional computational effort. It has further been demonstrated that the method can be cast into the form of conventional SEA with the addition of non-direct coupling loss factors, which provides a link with earlier work and with other enhancements of SEA. Although only plate structures have been considered here, the method is applicable to the same wide range of structures as SEA.

References

Bercin, A. N. 1993 High frequency vibration analysis of plate structures. Ph.D. thesis, Cranfield Institute of Technology, U.K.

Blakemore, M., Jeffries, B., Myers, R. J. M. & Woodhouse, J. 1990 Exploring statistical energy analysis on a cylindrical structure. *Proc. Inst. Acoust.* **12**, 587–593.

Cremer, L., Heckl, M. & Ungar, E. E. 1988 *Structure-borne sound*, 2nd edn. Berlin: Springer-Verlag.

Dowell, E. H. & Kubota, Y. 1985 Asymptotic modal analysis and statistical energy analysis of dynamical systems. *J. Appl. Mech.* **52**, 949–957.

Kubota, Y. & Dowell, E. H. 1986 Experimental investigation of asymptotic modal analysis for a rectangular plate. *J. Sound Vib.* **106**, 203–216.

Langley, R. S. 1989*a* Application of the dynamic stiffness method to the free and forced vibrations of aircraft panels. *J. Sound Vib.* **135**, 319–331.

Langley, R. S. 1989*b* A general derivation of the statistical energy analysis equations for coupled dynamic systems. *J. Sound Vib.* **135**, 499–508.

Langley, R. S. 1990 A derivation of the coupling loss factors used in statistical energy analysis. *J. Sound Vib.* **141**, 207–219.

Langley, R. S. 1992 A wave intensity technique for the analysis of high frequency vibrations. *J. Sound Vib.* **159**, 483–502.

Langley, R. S. & Heron, K. H. 1990 Elastic wave transmission through plate/beam junctions. *J. Sound Vib.* **143**, 241–253.

Lyon, R. H. 1975 *Statistical energy analysis of dynamical systems: theory and applications.* Cambridge, Massachusetts: MIT Press.

Zienkiewicz, O. C. 1977 *The finite element method.* London: McGraw-Hill.

Advanced statistical energy analysis

By K. H. Heron
Defence Research Agency, Farnborough, Hampshire GU14 6TD, U.K.

A high-frequency theory (advanced statistical energy analysis (ASEA)) is developed which takes account of the mechanism of tunnelling and uses a ray theory approach to track the power flowing around a plate or a beam network and then uses statistical energy analysis (SEA) to take care of any residual power. ASEA divides the energy of each sub-system into energy that is freely available for transfer to other sub-systems and energy that is fixed within the sub-system. The theory allows for coupling between sub-systems that are physically separate and can be interpreted as a series of mathematical models, the first of which is identical to standard SEA and subsequent higher order models are convergent on an accurate prediction. Using a structural assembly of six rods as an example, ASEA is shown to converge onto the exact results, whereas SEA is shown to overpredict by up to 60 dB.

1. Introduction

Statistical energy analysis (SEA) has been successfully applied to many noise and vibration problems. In particular SEA has become very useful as a framework for interpreting a vibro-acoustic data base. SEA often leads to a better understanding of the problem and SEA can point the way to practical solutions. However, when used as a purely predictive theory, without the recourse to measured data, SEA has not been universally successful. Nevertheless in some cases it has been very successful, for example when used to model the interaction between the noise in a room and its vibrating walls, but when applied to complex structural assemblies SEA predictions have often exhibited errors. These errors have been thought due to the fact that plates and beams are usually strongly coupled and one of the assumptions within standard SEA theory (see, for example, Lyon 1984) is that all couplings are weak. However, Keane & Price (1987) conclude that this assumption should be replaced by the necessity that no individual mode within a given sub-system should dominate the overall response of that sub-system, and this requirement can be met either by assuming weak coupling or by assuming the presence of many interacting modes. Furthermore, if SEA theory is developed using the wave approach rather than the modal approach this weak coupling assumption does not appear to be required (see, for example, Heron 1990).

In this paper we postulate that the errors that sometimes occur when predictive SEA is applied to complex structural assemblies are mainly due to an as yet unmodelled power transport mechanism. This 'tunnelling' mechanism conceptually occurs when direct coupling exists between two SEA sub-systems that are physically separated from each other by other SEA sub-systems. This mechanism of indirect coupling must not be confused with the power transport mechanism by which plate in-plane motion can couple physically separate bending motions; this latter phenomenon is fully modelled by existing SEA theory provided the in-plane sub-systems are included in the model.

A very simple form of tunnelling is associated with the non-resonant acoustic transmission through a plate and is already included in existing SEA theory (see, for example, Price & Crocker 1970; Leppington *et al.* 1987). However, this special case is mainly a function of the change in the dimension between the plate and the adjacent rooms and is not the concern of this paper.

Standard predictive SEA assumes zero coupling between the end plates of, for example, an in-plane assembly of three in-line plates. In this paper we develop a theory that allows for all sub-systems to be coupled to each other. Unlike for the simpler case of non-resonant acoustic transmission through a plate, we would intuitively expect this new theory to produce coupling loss factors that are not only a function of the power transmission coefficients across the various intervening line junctions, but are also dependent on the geometry and damping of the intervening plates.

In the following sections this theory is developed both for beam and plate networks and for want of a better name we will subsequently refer to this theory as advanced SEA or simply ASEA. Fundamentally it uses a ray theory approach to track the power flowing around the network and then uses standard SEA to take care of any residual power.

2. Free and fixed energy

Now all deterministic theories (modal, analytic, FEM, etc.) use field variables such as displacement and pressure and they must therefore include phase in the model, and the very essence of a high frequency model is the simplification associated with ignoring these phase effects. It is not just the need for computational efficiency that drives us to this assumption, but as Hodges & Woodhouse (1986) point out as we move to higher frequencies any deterministic approach becomes increasingly sensitive to the details of the physical structure under investigation, to such an extent that the results will be influenced by the deviations from the ideal design that inevitably occur in construction and such deviations are unknown. Thus all such deterministic approaches are rejected in this paper without further consideration; power accounting and the use of the sub-system energies as the field variables are the mainstays of SEA, and ASEA will be developed using the same philosophy.

The tunnelling phenomenon that we are attempting to model is associated with the transport of power, from sub-system 1 to sub-system 3 via an intervening sub-system 2 without at the same time inducing any increase in the 'free energy' of the intervening sub-system, and we must now consider what we mean by free energy.

With free energy we mean that part of the total sub-system energy that is available for transport to other sub-systems. In standard SEA all sub-system energy is free energy. Conversely the fixed energy of a sub-system is that part of the total sub-system energy that is not available for transport to other sub-systems. This postulates that the total energy of a SEA sub-system can be partitioned into a free and a fixed part is fundamental to ASEA theory.

Returning to the three in-line plate assembly example, we can now consider the following power flow mechanism. Free power associated with the free energy of sub-system 1 strikes the line junction between plate 1 and plate 2, this causes some power to transmit into plate 2 and as this power transports across plate 2 it will decrease in magnitude due to the damping mechanisms of plate 2. It is this loss of power that is self evidently not available for further transport duties and must be accounted for by a fixed energy field within plate 2. Finally some part of this transported power will strike the line junction between plates 2 and 3 where it will cause power injection into plate 3, and at this level of complexity such power will feed into the free energy field of plate 3.

3. SEA basics

First we find it helpful to rewrite the standard SEA matrix equation in a more convenient form for subsequent extension to ASEA such that

$$Ae = P - Me, \tag{3.1}$$

where $\boldsymbol{e}$ is a column vector of SEA modal energies, $\boldsymbol{P}$ is a column vector of input powers, M is a diagonal matrix of modal overlap factors, and A is a matrix of coupling loss factors. That is

$$M_{ii} = \omega n_i \eta_i, \tag{3.2}$$

where ω is the frequency, n_i is the modal density of sub-system i and η_i is the energy loss factor for sub-system i; furthermore for a three sub-system model we have

$$A = \omega \begin{bmatrix} n_1\eta_{12} + n_1\eta_{13} & -n_2\eta_{21} & -n_3\eta_{31} \\ -n_1\eta_{12} & n_2\eta_{21} + n_2\eta_{23} & -n_3\eta_{32} \\ -n_1\eta_{13} & -n_2\eta_{23} & n_3\eta_{31} + n_3\eta_{32} \end{bmatrix}, \tag{3.3}$$

where η_{ij} is the usual SEA coupling loss factor.

Of course the more usual SEA matrix equation can be recovered by combining the A and the M matrices in (3.1). The reason for the above formulation will become apparent as we develop ASEA theory, but for now it is worth noting that each of the three terms in (3.1) have a clear physical meaning; the left-hand side term incorporates all the power transport and coupling effects and the two right-hand side terms model all the power sources and all the power sinks respectively.

Furthermore, if all the equations in (3.1) are added together we have, by power balance, that the sum of all the right-hand side terms is zero, and this is true for all possible $\boldsymbol{P}$ and thus for all possible $\boldsymbol{e}$. Hence each individual column of A must always sum to zero, which is of course a trivial deduction from SEA. Indeed, assuming SEA reciprocity, A is a symmetric matrix and thus each individual row of A must also sum to zero. However, it is important to note that this row sum rule is a consequence of power balance and SEA reciprocity whereas the column sum rule is solely a consequence of the much more fundamental requirement of power balance.

4. ASEA basics

In developing ASEA theory we will, as described above, split the total energy field within each sub-system into two parts, a free energy field with a modal energy of e, and a fixed energy field with an 'equivalent' modal energy of d. The term 'modal energy' is used because of its historic link with classical SEA theory, however the reader might find it easier to think of the modal energy as a measure of the energy density of a sub-system with which it is closely related for sub-systems made up of simple beams, plates or rooms.

Using the column vectors $\boldsymbol{e}$ and $\boldsymbol{d}$ as the field variables, ASEA can be encapsulated by the following two matrix equations

$$\underset{\text{free power to free power transfer}}{A\boldsymbol{e}} = \underset{\text{free power input}}{\boldsymbol{P}} - \underset{\text{free power lost}}{M\boldsymbol{e}}, \tag{4.1}$$

$$\underset{\text{free power to fixed power transfer}}{B\boldsymbol{e}} = \underset{\text{fixed power input}}{\boldsymbol{Q}} - \underset{\text{fixed power lost}}{M\boldsymbol{d}}, \tag{4.2}$$

and to understand better these equations we have attached a physical description to each of the terms. The above equations form the basis for ASEA theory and this paper is mainly concerned with the calculation procedure for the A and the B matrices.

It may be thought that the somewhat arbitrary use of M in the second equation involves an assumption but this is not so since we have yet to specify the precise definition of B and $\boldsymbol{Q}$, and the requirement to conform with equation (4.2) creates those definitions.

Once A, B, $\boldsymbol{P}$ and $\boldsymbol{Q}$ are known the responses can be calculated from $\boldsymbol{e}+\boldsymbol{d}$, using exactly the same procedures that we currently use when calculating SEA responses from $\boldsymbol{e}$. It should be noted that the A matrix of ASEA theory is not the same as the A matrix of standard SEA theory.

From equations (4.1) and (4.2), $\boldsymbol{e}+\boldsymbol{d}$ is given by

$$\boldsymbol{e}+\boldsymbol{d} = M^{-1}(\boldsymbol{Q}+\boldsymbol{R}), \tag{4.3}$$

where

$$\boldsymbol{R} = (M-B)(M+A)^{-1}\boldsymbol{P}. \tag{4.4}$$

Now for the classical excitation of 'rain on the roof' $\boldsymbol{Q}$ is zero, and with this simplification equation (4.3) can be rewritten as

$$(M+A)(M-B)^{-1}M(\boldsymbol{e}+\boldsymbol{d}) = \boldsymbol{P}, \tag{4.5}$$

and this equation can be considered to be the 'equivalent' standard SEA matrix equation such that if

$$A_{\text{sea}}\boldsymbol{e}_{\text{sea}} = \boldsymbol{P}, \tag{4.6}$$

then

$$A_{\text{sea}} = (M+A)(M-B)^{-1}M, \tag{4.7}$$

and

$$\boldsymbol{e}_{\text{sea}} = \boldsymbol{e}+\boldsymbol{d}. \tag{4.8}$$

Finally by applying the same power balancing argument of §3 we can easily deduce the important property that each individual column of $A+B$ must always sum to zero.

5. ASEA and beam networks

In a beam network each beam will consist of four sub-systems associated with its two bending wavetypes, its compressional wavetype and its torsional wavetype. In this section, for clarity of presentation, we will only consider a network of rods with each rod having only one wavetype. Provided we allow for this one wavetype to be conceptually of any type, for example by not assuming that the group velocity is equal to the phase velocity, then the extension to a beam network is straightforward.

Consider now the free energy field of rod j, represented by its modal energy e_j. Then the total free energy of this rod, E_j, is given by

$$E_j = n_j e_j, \tag{5.1}$$

and the energy density of this free energy is E_j divided by L_j, where L_j is the length of rod j. Now by assuming that this energy field is made up of equal amounts of incoherent power, P_j, flowing both from left to right and from right to left along the rod (equivalent to the random incidence assumption in two- and three-dimensional sub-systems) we have

$$E_j/L_j = 2P_j/c_{gj}, \tag{5.2}$$

where c_{gj} is the group velocity of rod j.

Furthermore since for all one-dimensional sub-systems

$$n_j = L_j/\pi c_{gj}, \tag{5.3}$$

we can combine equation (5.1) and equation (5.2) to obtain the standard SEA result that

$$P_j = e_j/2\pi. \tag{5.4}$$

Thus for unit modal energy the power available at each end of rod j, P_{aj} say, for potential transportation to the other rods, is simply $\frac{1}{2}\pi$.

We can now proceed with the calculation of the elements of the matrices A and B. Initially all these are set to zero and the calculation is based on using the elements of these matrices as accumulators. We start by taking a particular end of a particular rod and ultimately repeat the calculation for both ends of every rod.

The power available per unit modal energy P_{aj} at this particular end of rod j will conceptually be all transferred from rod j, and thus P_{aj} must now be added to element (j,j) of matrix A; add rather than subtract because the transfer terms have been conventionally placed on the left-hand side of equation (4.1) and equation (4.2).

Now we take this available power, P_{aj} say, and multiply it by the appropriate transmission or reflection coefficient. This is then the power at the connected end of a particular receiving rod, rod i say, and this power is now ready for transportation across this rod; rod i can be the same rod as rod j to take care of the reflected wave and indeed the following calculations must be performed for all rods connected to the chosen end of rod j including rod j itself. This start power, P_{si} say, is thus given by

$$P_{si} = \tau_{ij} P_{aj}, \tag{5.5}$$

where τ_{ij} is the power transmission coefficient for power flowing from rod j to rod i.

It is worth keeping in mind at this point the standard SEA theory which would proceed in the following manner

$$\omega n_j \eta_{ji} = P_{si} = \tau_{ij} P_{aj} = \tau_{ij}/2\pi, \tag{5.6}$$

and thus

$$\eta_{ji} = \tau_{ij}/2\pi\omega n_j. \tag{5.7}$$

Returning to ASEA theory, power will flow across rod i and will decay as it does so with the exponential factor

$$\exp(-\omega\eta_i L_i/c_{gi}) = \exp(-\pi M_i), \tag{5.8}$$

where M_i is the modal overlap factor of rod i. Thus

$$P_{ei} = \exp(-\pi M_i) P_{si}, \tag{5.9}$$

where P_{ei} is the power striking the far end of rod i. The power lost during this crossing, P_{li} say, is given by

$$P_{li} = P_{si} - P_{ei}. \tag{5.10}$$

This lost power must now be subtracted from element (i,j) of matrix B; matrix B rather than matrix A since this power is self evidently unavailable for further transport duties. On the other hand, P_{ei} is available for further transport duties, and indeed we can continue the calculation from equation (5.5) using P_{ei} rather than P_{aj}. Of course within this cycle of the calculation we can only modify column j of either matrix A or matrix B since all of the initial available power comes from rod j. This whole process can be stopped at any stage and having stopped any remaining power, P_{si} say, must then be subtracted from element (i,j) of matrix A. This latter is essential to maintain power balance and conceptually uses a standard SEA approach to sweep up and account for the residual power P_{si}; it also ensures that all the columns of $A+B$ sum to zero as required by power balance.

6. ASEA and plate networks

The above theory can be extended to plate networks although its actual implementation could well turn out to be computationally expensive, as compared with standard SEA. However, ASEA plate theory will hopefully guarantee an accurate prediction and the fact that it may not become a practical tool because of the computational load should not deter us from its development. Its use as a tool for the validation of more approximate theories is very important because no accurate high frequency theory exists for general structural assemblies.

Whereas with rods we calculated the A and the B matrices by starting with a particular end of a particular rod and with beams we would start with a particular end of a particular beam and with a particular wavetype, with plates we must start with a particular edge of a particular plate and not only with a particular wavetype but also with a particular incidence angle at the chosen edge. In standard SEA the eventual integral over all angles of incidence is carried out implicitly within the model such that the formula for an SEA plate to plate coupling loss factor is a function of the random incidence transmission coefficient as given below in equation (6.6). In ASEA we can only perform the integral over all possible angles of incidence, 180°, at the end of the A and B calculation; although by converting this integral into a suitably weighted sum we can easily incorporate it into the calculation procedure. Unfortunately line junction transmission coefficients tend to vary a lot with angle of incidence due mainly to the complex interaction effects of the various wavetypes and it is often necessary to perform these calculations over many angles of incidence: typically at every integer degree.

For a random diffuse energy field in sub-system j of a plate the intensity, I_j say, is given by

$$I_j = e_j k_j / 4\pi, \tag{6.1}$$

where k_j and e_j are the wavenumber and modal energy respectively of the wavetype associated with sub-system j. The power per unit modal energy striking an edge of length L at a grazing angle of incidence ϕ_j is thus

$$P_{aj} = L k_j \sin(\phi_j)/4\pi, \tag{6.2}$$

and as before this must now be added to element (j,j) of matrix A.

We set

$$P_{si} = \tau_{ij}(\phi_j) P_{aj}, \tag{6.3}$$

however, τ_{ij} is now a function of ϕ_j and the transmitted wave angle has to be calculated using trace wavenumber matching such that

$$k_i \cos(\phi_i) = k_j \cos(\phi_j). \tag{6.4}$$

Again at this point it is worth keeping in mind the standard SEA theory which for plates proceeds as follows

$$\omega n_j \eta_{ji} = \pi^{-1} \int_0^{\pi} P_{si}\, \mathrm{d}\phi_j, \tag{6.5}$$

and thus

$$\eta_{ji} = L k_j \hat{\tau}_{ij} / 2\pi^2 \omega n_j, \tag{6.6}$$

where the random incidence transmission coefficient is given by

$$\hat{\tau}_{ij} = \int_0^{\pi} \tau_{ij}(\phi_j) \sin(\phi_j)\, \mathrm{d}\phi_j. \tag{6.7}$$

Returning to ASEA theory, geometric calculations must now be made to track the wave as it is transported across sub-system i. This can result in more than one edge of the plate supporting sub-system i being illuminated and furthermore an illuminated edge need not be illuminated along its entire length; both of these effects must be calculated.

The damping factor, equivalent to the factor $\mathrm{e}^{-\pi M}$ of equation (5.8), is also more complicated here. Different parts of the wave will travel different distances, however for polygon shaped plates a damping factor averaged over all possible path lengths between two edges can be used and this is given by

$$\frac{(\mathrm{e}^{-b\kappa} - \mathrm{e}^{-a\kappa})}{(a\kappa - b\kappa)} = D, \tag{6.8}$$

where

$$\kappa = \omega \eta_i / c_{gi} = 2\pi M_i / A_i k_i, \tag{6.9}$$

and where a and b are the maximum and minimum path lengths.

Finally

$$P_{ei} = D P_{si} \tag{6.10}$$

and

$$P_{li} = P_{si} - P_{ei}, \tag{6.11}$$

and P_{li} is subtracted from element (i,j) of matrix B as before.

7. Comparison with analytical results

ASEA produces a different result dependent on the number of transfers of power across a sub-system that we are modelling. This number which is also one less than the number of junctions crossed we will call the ASEA level number, and with a level number of zero ASEA always produces results identical to standard SEA since both B and $\boldsymbol{d}$ are then zero. Advanced SEA can thus be thought of as a series of approximations,

$$\mathrm{ASEA}_0\ (\equiv \mathrm{SEA}), \mathrm{ASEA}_1, \mathrm{ASEA}_2, \mathrm{ASEA}_3, \ldots, \tag{7.1}$$

with the expectation that this series converges on the required result.

It is important to understand why we have this clear expectation that if the series (7.1) converges at all it must converge onto the 'correct' result; correct in the sense of giving the best high frequency result possible.

Consider the calculation procedure for ASEA with a very large level number; the level number chosen to be so large as to cause the A matrix to be effectively zero. Then the ASEA calculation procedure is nothing more than ray tracing with all phase related effects ignored, or in other words simple power flow analysis. But unless we want to encroach on the low frequency deterministic domain, any high frequency theory must at least make the assumption, explicitly or implicitly, that all phase effects be ignored. Now with this assumption, and this assumption alone, we can deduce ASEA for an infinite level number. (Self evidently this would also be true for a simple power flow analysis, the subtle difference is that ASEA hopefully converges much faster due to the different treatment of the 'remainder' terms, which are ignored in a simple power flow analysis but are injected into a SEA procedure whose results are added to the truncated power flow analysis during an ASEA calculation.) Thus we fully expect that, if ASEA converges at all, and if an accurate high frequency theory exists at all, ASEA will converge onto the best theoretical result possible.

To show this convergence for a particular case we have chosen a very simple assembly consisting of six different rods all in a line. In principle an assembly of plates could equally well have been chosen; however, exact results are extremely difficult to compute for plate assemblies at high frequencies and thus we have chosen an assembly of rods. The inline configuration has been deliberately chosen to highlight the errors in a simple SEA calculation and the subsequent correction of these errors by ASEA. The inability of SEA to predict such a contrived configuration is understandable and does not detract from the usefulness of SEA when applied to more realistic structures, but it should be considered as a warning that the accuracy of SEA is structure dependent.

The coupling between the rods is such that conceptually the whole structure could be made from a single piece of material with the far ends of the chain left unsupported or free. The rod material is such that its longitudinal phase, or group, velocity is 5000 m s^{-1}. The six rods are of lengths 23, 28, 25, 24, 29 and 21 m and their cross-sectional areas are such that their mass per unit lengths are 1, 10, 3, 7, 8 and 2 kg m^{-1} respectively. An energy damping value of 2% was chosen for the SEA modelling, and viscous damping with an equivalent critical damping ratio of 0.01 chosen for the exact model. The structure was always driven with a unit force on the first rod.

The exact results were calculated by Keane (1992; personal communication) and form a full deterministic analysis for point excitation, they are based on calculating the power flow across the assumed point connections between the rods for a given unit point force excitation on the drive rod. These response data were then numerically averaged over all excitation positions on the drive rod, rod 1, and over all frequencies within the chosen frequency bandwidth of 50 Hz.

Figure 1*a–d* shows the results for the averaged response on the four rods furthest from the drive rod; the results for rods 1 and 2 are not shown because SEA and all levels of ASEA lie very close to the exact results for these rods. All the displayed responses have been normalized to unit mean square response velocity at the drive point on rod 1.

As can be seen SEA, or equivalently ASEA_0, is not an adequate model at the higher frequencies; at 10 kHz SEA over predicts the response of rod 6 by over 60 dB. On the other hand, as expected, ASEA always predicts accurately provided we are willing to calculate to a high enough level number. For a chain of rods driven at one end the rule of convergence appears to be that the ASEA level number should be at least the rod number minus two. This is not so surprising a result since such a level number ensures a direct coupling exist in the ASEA model between the drive rod and the response rod. The convergence of ASEA is not necessarily monotonic with level number as can be seen in figure 1*d*, where ASEA_2 gives a slightly better result than ASEA_3.

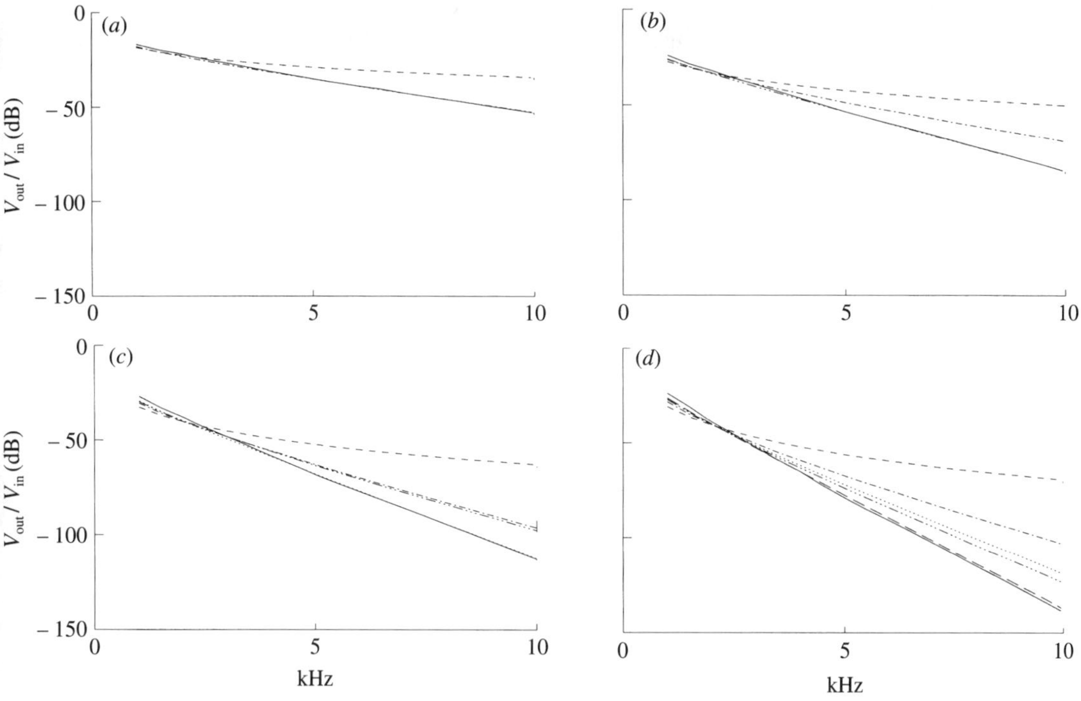

Figure 1. Response of rod 3(*a*), 4(*b*), 5(*c*) and 6(*d*). ——, Exact result; ----, $ASEA_0$ prediction; —·—, $ASEA_1$ prediction; —···—, $ASEA_2$ prediction; ....., $ASEA_3$ prediction; ——, $ASEA_4$ prediction.

8. Conclusions

A high frequency theory (ASEA) has been presented that takes account of the mechanism of tunnelling. This mechanism which requires the introduction of coupling between SEA sub-systems that are physically separate is modelled by creating a new set of basic ASEA equations and dividing the energy of a sub-system into energy that is freely available for transfer to other sub-systems and energy that is fixed within the sub-system.

These equations are presented and an attempt has been made to give their component parts physical meaning. The calculation procedure is presented for modelling either a general beam network or a general plate network. ASEA is interpreted as a series of mathematical models, the first of which is identical to standard SEA and subsequent higher order models are convergent on the desired result.

Using a structural assembly of six rods as an example, ASEA converges onto the exact results whereas SEA is shown to overpredict by up to 60 kB.

I thank the U.K. Ministry of Defence for the financial support given to this work programme.

References

Heron, K. H. 1990 The development of a wave approach to statistical energy analysis. In *Proc. Inter-noise*. Gothenburg, Sweden.

Hodges, C. H. & Woodhouse, J. 1986 *Rep. Prog. Phys.* **49**, 107–170.

Keane, A. J. 1992 *Proc. R. Soc. Lond.* A **436**, 537–568.

Keane, A. J. & Price, W. G. 1987 *J. Sound Vib.* **117**, 363–386.

Leppington, F. G., Heron, K. H., Broadbent, E. G. & Mead, S. M. 1987 *Proc. R. Soc. Lond.* A **412**, 309–337.

Lyon, R. H. 1984 *Statistical energy analysis of dynamic systems: theory and application*. Cambridge: MIT Press.

Price, A. J. & Crocker, M. J. 1970 *J. Acoust. Soc. Am.* **473**, 683–693.

Vibration transmission through symmetric resonant couplings

By D. J. Allwright[1], M. Blakemore[1], P. R. Brazier-Smith[1] and J. Woodhouse[2]

[1]*SAIC S&E Ltd, Poseidon House, Castle Park, Cambridge CB3 0RD, U.K.*
[2]*Cambridge University Engineering Department, Trumpington Street, Cambridge CB2 1PZ, U.K.*

The transmission of vibration through a symmetric junction is considered. The problem is introduced using a stretched string with a general point attachment, and then a result is derived which encapsulates the important aspects of the transmission behaviour for a wider class of systems. These are systems that consist of two semi-infinite sections of identical, one-dimensional structure having only one propagating wavetype (but any number of evanescent ones), joined through any linear system that satisfies a condition of symmetry. For such systems, it is shown that there will in general be a set of frequencies of perfect transmission and perfect reflection, in a number and pattern which can be described in terms of the behaviour of the junction alone. Representative examples are presented, based on the behaviour of bending beams and thin circular cylinders with attached structures providing wave reflection. The implications of this result are explored for SEA coupling loss factors, and for the interpretation of SEA model predictions when such resonant coupling structures are present.

1. Introduction

Fundamental to many problems in noise and vibration control is the understanding of vibration transmission between two systems across a junction of some kind. The level of transmitted vibration, and its distribution in space and frequency, will be governed by the detailed design of the systems and the junction. There are many techniques, theoretical and experimental, for investigating such transmission behaviour. At low frequencies the whole system is likely to exhibit low modal overlap, so that deterministic analysis using idealized theoretical models, finite element analysis, or experimental modal analysis can commonly be used to good effect. Higher in frequency, especially if the modal overlap becomes significant, such treatment becomes very difficult to carry through with acceptable accuracy, and in any case may be of dubious value. One may learn more from an approximate analysis of a stochastic nature, such as statistical room acoustics or statistical energy analysis (SEA) (see, for example, Lyon 1975; Hodges & Woodhouse 1986).

This paper addresses a class of problems that are often encountered in practice, and which present difficulties for both these styles of analysis. It concerns junction structures that have internal degrees of freedom, so that their dynamics must be taken into account in the transmission calculation. If the frequency range of interest is not too high in the modal series of the subsystems coupled through this junction, then a deterministic analysis of the whole coupled system may be possible which incorporates the junction behaviour. But at higher frequencies the modal overlap may become high in the subsystems so that a statistical analysis is indicated, but the modal density of the junction structure on its own can still be quite low, so that it may not be appropriate to include it as a third energy-storing subsystem in, for example, a SEA model.

A common starting point for analysing such a system would be to consider the wave reflection/transmission problem in which the two subsystems were regarded as semi-infinite (see, for example, Lyon 1975; Cremer *et al.* 1973). The transmission coefficient across the junction can be calculated as a function of frequency, and if the subsystems are two- or three-dimensional, as a function of angle of incidence at the junction. If a SEA model is required, the coupling loss factor would then be calculated from this transmission coefficient by an averaging procedure (Lyon 1975). This procedure can be applied, at least in principle, to any combination of subsystems and junction. The existence of internal resonances in the junction does not invalidate the approach, but it has significant implications for the interpretation of the results of any modelling. Such internal resonances produce strong frequency dependence of the transmission coefficient, which has immediate consequences for a deterministic study, and rather less obvious ones for the results of a statistical model. The nature and implications of this frequency dependence form the main subject of this study.

2. Wave transmission past a point attachment on a stretched string

Consider a stretched string of tension P and line density m, having a constraint attached at the point $x = 0$, in the form of a linear system which presents a frequency-dependent (velocity) admittance $Y(\omega)$ to the transverse motion of the string. Assume a transverse displacement field

$$y(x,t) = e^{-i\omega t}[e^{ikx} + R\,e^{-ikx}] \quad (x \leqslant 0), \quad y(x,t) = T e^{-i\omega t} e^{ikx} \quad (x \geqslant 0) \tag{2.1}$$

on the string, where $k = \omega(m/P)^{\frac{1}{2}}$. Enforcing the constraint condition at $x = 0$, the transmission coefficient T and reflection coefficient R are readily shown to be

$$T = \frac{2Y}{2Y+Y_0} \quad \text{and} \quad R = \frac{-Y_0}{2Y+Y_0}, \tag{2.2}$$

where $Y_0 = (Pm)^{-\frac{1}{2}}$ is the wave admittance of the string.

We discuss exclusively conservative constraint systems, so that $Y(\omega)$ is purely imaginary at all frequencies. It follows that

$$T/R = -2Y/Y_0 \tag{2.3}$$

is purely imaginary, and that

$$|T|^2 + |R|^2 = \frac{|2Y|^2 + |Y_0|^2}{|2Y+Y_0|^2} = 1 \tag{2.4}$$

as expected.

For any frequency such that $Y \rightarrow \infty$, there is perfect transmission of transverse waves past the constraint ($T = 1$, $R = 0$). Conversely, at any frequency such that $Y \rightarrow 0$ there is perfect reflection ($T = 0$, $R = -1$). The former case occurs trivially in the absence of any constraint, or at a resonance frequency of the constraining system. The latter case occurs if the constraint takes the form of a fixed point, or at

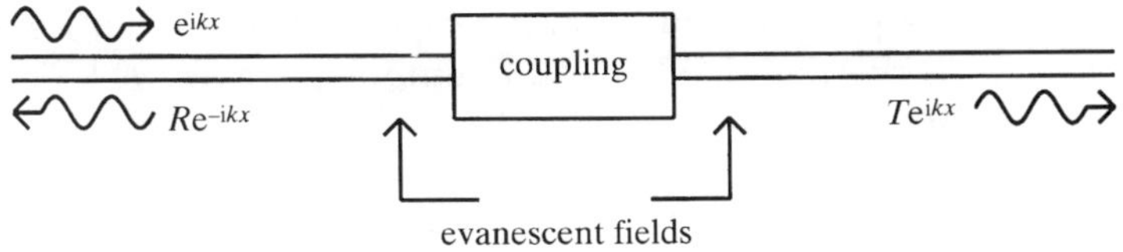

Figure 1. Sketch of two-semi-infinite one-dimensional wave-bearing systems joined by a symmetric coupling.

an antiresonance of the constraining system. Since $Y(\omega)$ is the driving-point admittance of a linear system, resonance and antiresonance frequencies alternate (see, for example, Skudrzyk 1980). Thus frequencies of perfect transmission and perfect reflection alternate, both occurring on average at the modal density of the constraining system alone.

The function $Y(\omega)$ may be written in terms of the eigenvalues and normal modes of the constraining system (Hodges & Woodhouse 1986)

$$Y(\omega) = -\sum_n \frac{\mathrm{i}\omega u_n^2}{\omega_n^2 - \omega^2}, \tag{2.5}$$

where u_n is the nth mass-normalized mode shape (evaluated at the point of attachment to the string) and ω_n its frequency. This result makes it easy to investigate the width of the transmission peaks. Near the nth modal frequency of the constraining system, assuming low modal overlap, the admittance will be well approximated by one term from the summation of (2.5), so that

$$T \approx \frac{2\mathrm{i}\omega u_n^2}{2\mathrm{i}\omega u_n^2 - Y_0(\omega_n^2 - \omega^2)}. \tag{2.6}$$

This has the familiar form of the response of a damped single-degree-of-freedom system to forcing, so that we may write down the 'half-power bandwidth' of the perfect-transmission peak in $T(\omega)$:

$$\Delta_n \approx 2u_n^2(0)/Y_0. \tag{2.7}$$

This bandwidth has its origin in the radiation damping of the constraint mode induced by connecting it to the semi-infinite strings.

3. A symmetric constraint on a one-dimensional system

Much of the behaviour seen in this simple example can be generalized to a class of one-dimensional wave-bearing systems, on which an attached structure or other inhomogeneity produces some reflection of waves. The system to be considered is shown schematically in figure 1. Two semi-infinite homogeneous sections capable of supporting a single propagating wavetype are connected through an intermediate system, which may act at a point or be of finite extent in the x direction. The only stipulations on this junction are that it be conservative and symmetric with respect to the transformation $x \rightarrow -x$.

For an incident wave from the left, reflection and transmission coefficients R and T may be defined by assuming displacement fields as shown in figure 1. From the assumption of symmetry, the same coefficients will govern reflection and transmission of a wave incident from the right. Now consider symmetric excitation of the system, with waves of equal amplitude and phase incident from both right and left. After interaction with the junction structure, outgoing waves $(R+T)\,e^{-i(\omega t \pm kx)}$ will travel to the left (+ sign) and right (− sign). Evanescent waves will also be produced near the junction, decaying away from it symmetrically on both sides. Internal motions of the junction structure will be excited, but only those which are symmetric under $x \to -x$.

Under this symmetric excitation, it is plain that there is no net energy flux passing any point in the system. Considering the combined wavefield beyond the reach of the evanescent fields, it follows that $|R+T| = 1$. So let

$$R+T = e^{2i\theta_+}, \tag{3.1}$$

where θ_+ is a function of frequency, which is well-defined mod π and can be chosen continuous. The displacements in the wavefield, on the left say, away from the evanescent parts are then

$$e^{-i(\omega t - kx)} + e^{2i\theta_+}\, e^{-i(\omega t + kx)} = 2e^{-i(\omega t - \theta_+)} \cos(kx - \theta_+) \tag{3.2}$$

so that they are in phase everywhere.

Now choose a symmetric pair of points on the left and the right in the far wavefield, at an antinodal point of this standing wave. Formally, we can join these two points by imposing periodic boundary conditions, without in any way changing the displacement fields. The result is a finite system that satisfies the condition of phase closure at the artificially joined point. But this is one standard way of determining normal mode frequencies of a finite system (see, for example, Cremer *et al.* 1973 §II, 4). So we may deduce that all other generalized coordinates, including those describing the evanescent fields and the internal degrees of freedom of the joint structure, are also moving in the same phase. Thus we may regard θ_+ as the phase of a transfer function, taking the symmetric pair of incoming waves as input and the response of one of the generalized coordinates describing the joint structure as output. But the behaviour of the phase of forced response of a resonant structure is very familiar: it increases by π for each mode which is excited. In this case, these are the modes of the junction which are symmetric under $x \to -x$. Of course, we ignore any modes which are uncoupled to the wave motion. We also assume that the modes are non-degenerate.

This argument may be repeated for the case of an antisymmetric pair of incoming waves (equal amplitude but opposite phase in the far wavefield). Analogously to (3.1), we find

$$R-T = e^{2i\theta_-}, \tag{3.3}$$

say, where θ_- changes by π each time the frequency passes through a mode of the junction structure which involves motion which is antisymmetric under $x \to -x$. Combining the two cases, we may deduce

$$\frac{R+T}{R-T} = e^{2i(\theta_+ - \theta_-)}, \tag{3.4}$$

from which it follows that

$$T/R = i \tan(\theta_+ - \theta_-). \tag{3.5}$$

This shows that the result displayed for the string with a point attachment in (2.3) is quite general: the transmitted and reflected waves are in quadrature for any system satisfying the symmetry conditions assumed here.

It also follows immediately from (3.5) that whenever $(\theta_+ - \theta_-) = n\pi$, $T = 0$ and there is perfect reflection of waves from the junction. Conversely, whenever $(\theta_+ - \theta_-) = (n + \frac{1}{2})\pi$, $R = 0$ and there is perfect transmission through the junction. Combining these results with the earlier remarks about the frequency dependence of θ_+ and θ_- yields a direct link between the resonances of the junction structure and the number and distribution of frequencies of perfect reflection and perfect transmission. Suppose first that the junction has only symmetric resonances, so that θ_- is constant. Then θ_+ is monotonically increasing, through π for each resonance, and there is a strict alternation of frequencies of perfect reflection and perfect transmission. This is the case illustrated in §2, since a point attachment to a stretched string could only exhibit symmetric modes of vibration (as a string cannot support a point moment applied to it). Similar behaviour would occur in the converse case, in which the junction structure allowed only antisymmetric modes.

The general case is rather more complicated. The behaviour depends on $\theta_+ - \theta_-$, so that there is scope for the influences of symmetric and antisymmetric junction modes to interact. If the modes are well separated (compared with the extent by which they are broadened by radiation damping), there will be no significant interaction. The phases θ_+ and θ_- each flip by π for each resonance, so that corresponding to each resonance of the junction structure alone we may expect to find one frequency of perfect transmission and one of perfect reflection. The precise frequencies at which these occur depend on how the junction mode is influenced by coupling to the evanescent and travelling waves of the wave-bearing systems. If the modes overlap, things are less clear-cut. If there is a difference in modal densities between symmetric and antisymmetric junction modes, then whatever happens there is an inexorable trend in $\theta_+ - \theta_-$ based on the difference of the two, and that sets a minimum density of frequencies of perfect reflection and perfect transmission. But they will in general occur more frequently than this minimum, as will be illustrated by examples in the next section.

4. Deterministic examples

(*a*) *Point-constrained bending beam*

The simplest system to exhibit the full range of behaviour revealed in the previous section is a bending beam (with line density m and bending stiffness B) with a point attachment that has resonances in both transverse and torsional motion (but no coupling between the two, to satisfy the assumption of symmetry). So suppose that at the point $x = 0$, a linear system is attached which presents a transverse admittance $Y_t(\omega)$ and a rotational admittance $Y_r(\omega)$. For an incident wave from the left, assume displacement fields

$$u(x,t) = e^{-i\omega t}[e^{ikx} + R\,e^{-ikx} + D\,e^{kx}] \quad (x \leqslant 0),$$
$$u(x,t) = e^{-i\omega t}[T\,e^{ikx} + F\,e^{-kx}] \quad (x \geqslant 0), \tag{4.1}$$

where $k = [m\omega^2/B]^{\frac{1}{4}}$. Imposing the four boundary conditions at $x = 0$ and solving the resulting simultaneous equations, the solution for the reflection and transmission coefficients may conveniently be written in the form

$$R + T = \frac{2(1-i) - h}{2(1-i) + ih}, \quad R - T = -\frac{2(1+i) + g}{2(1+i) + ig}, \tag{4.2}$$

where $$g = -i\omega/Y_r Bk, \quad h = -i\omega/Y_t Bk^3. \tag{4.3}$$

In terms of the formalism of §3, this yields

$$\theta_+ = \arctan(1-\tfrac{1}{2}h)-\tfrac{1}{4}\pi, \quad \theta_- = -\arctan(1+\tfrac{1}{2}g)-\tfrac{1}{4}\pi. \tag{4.4}$$

As expected, θ_+ depends only on the symmetric motion of the constraint, via the function h involving the transverse response, while θ_- depends only on the antisymmetric motion via the function g involving rotational response. Solving for the symmetric and antisymmetric combination of the evanescent fields yields

$$F+D = -2\sin\theta_+ \, e^{i\theta_+}, \quad F-D = 2\cos\theta_- \, e^{i\theta_-}. \tag{4.5}$$

These have the appropriate phases θ_+ and θ_- respectively, as predicted by the general argument in §3.

From (4.4) it is possible to deduce the behaviour of θ_+ and θ_- in various limiting cases. If the symmetric constraint is weak, then h will be small at most frequencies. For a strong constraint, on the other hand, h will generally be large. Similar remarks apply to the function g in respect of the strength of antisymmetric constraint. The symmetric and antisymmetric constraints do not necessarily have the same order of magnitude of strength, as will be seen in §4*b*. Now it is clear that when $h = 0$, $\theta_+ = 0$, and when $h \to \infty$, $\theta_+ = \frac{1}{4}\pi$. Similarly, when $g = 0$, $\theta_- = \frac{1}{2}\pi$, and when $g \to \infty$, $\theta_- = \frac{1}{4}\pi$. (All these phases are given mod π.) So for weak or strong constraints with the respective symmetries, the corresponding phases can be expected to lie close to these limiting values at most frequencies, flipping rapidly through π when the pattern of constraint resonances and antiresonances requires it. It follows that if both constraints are weak, then $|\theta_- - \theta_+| \approx \frac{1}{2}\pi$ and $|T| \approx 1$ at most frequencies as one would expect. Conversely if both constraints are strong, $|\theta_- - \theta_+| \approx 0$ so that $|T| \approx 0$ at most frequencies. Finally, if one constraint is weak and the other strong (an extreme case being a rigidly pinned point constraint), then $|\theta_- - \theta_+| \approx \frac{1}{4}\pi$, and $|T| \approx 1/\sqrt{2}$.

To see this behaviour in detail, it is convenient to work in terms of a specific example. Let the attached system be a finite section of another bending beam lying parallel to the infinite beam. This is assumed to be rigidly attached to the infinite beam at its centre point, and to have both ends free so that it acts as a double cantilever. The symmetric vibration modes of the finite beam will couple to the wave-bearing beam via a transverse force, and will govern the behaviour of θ_+. The antisymmetric modes have a nodal point at the beam centre, but will couple via a moment and will govern θ_-. The two sets of modes alternate on the unconstrained finite beam, of course. The system may be regarded as a multi-mode 'tuned absorber' attached to the infinite beam (although no damping is allowed in the system for the present purpose).

By allowing the attached beam to have different properties to those of the infinite beam, it is possible to investigate different régimes of strength of coupling. We will suppose that the finite beam has bending stiffness λB and line density λm. Small values of λ will correspond to weak constraint (for both symmetric and antisymmetric motion), while large values of λ will correspond to strong constraint. The functions θ_+ and θ_- can be readily computed, using the fact that the two admittance functions for this particular constraint system are

$$Y_t = \frac{i\omega}{2\lambda B k^3} \frac{[1+\cos kL \cosh kL]}{[\cosh kL \sin kL + \cos kL \sinh kL]} \tag{4.6}$$

and

$$Y_r = \frac{i\omega}{2\lambda B k} \frac{[1+\cos kL \cosh kL]}{[\cosh kL \sin kL - \sinh kL \cos kL]}, \tag{4.7}$$

where the attached beam is of length $2L$.

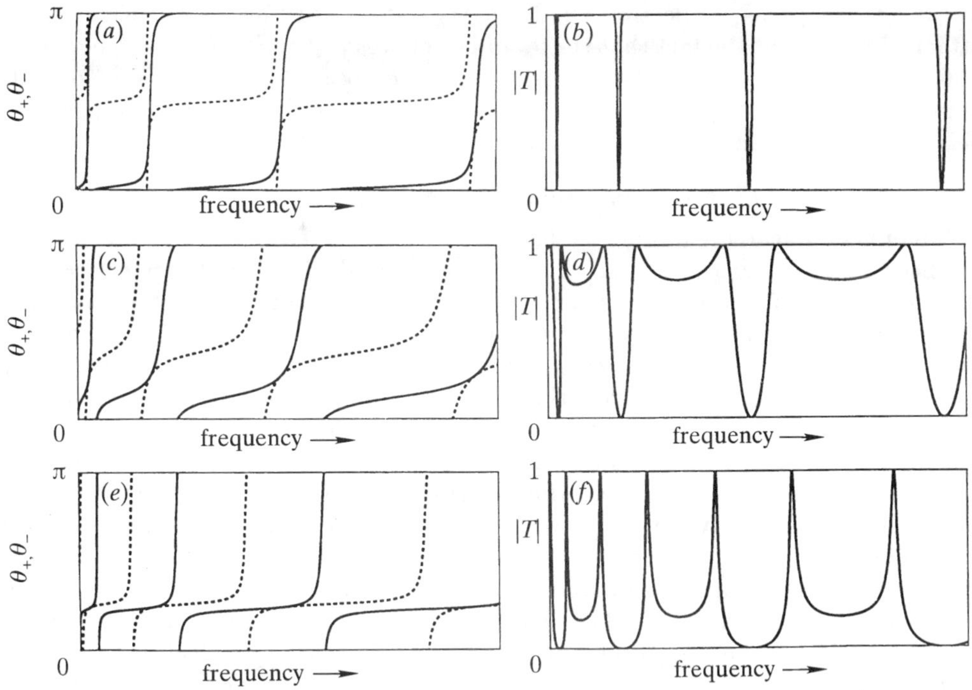

Figure 2. Phase angles θ_+ and θ_- (left-hand column) and transmission coefficients (right-hand column) plotted against frequency, for a symmetric free-free beam rigidly attached at its centre to an infinite beam. The attached beam has bending stiffness and line density λ times those of the infinite beam, where $\lambda = 0.1$ (top row), 1 (middle row) and 10 (bottom row).

Results are shown in figure 2*a*, *c*, *e*, for values $\lambda = 0.1$, 1 and 10. These illustrate the general characteristics expected from the preceding discussion. For $\lambda = 0.1$, θ_+ has plateaux around the value 0, while θ_- has them around $\frac{1}{2}\pi$. Conversely for $\lambda = 10$, the plateaux lie at $\frac{1}{4}\pi$ for both phase angles. For $\lambda = 1$ intermediate behaviour is seen, with less strongly marked plateaux. A similar sequence of behaviour would be expected from any other attached constraint system, as the strength of coupling was varied. The set of θ_+ curves all pass through the same values at the points where $\theta_+ = 0$ and $\frac{1}{4}\pi \pmod{\pi}$. This is because at the frequencies at which $h = 0$ or ∞, the multiplying factor governing the constraint strength does not matter. Similarly, the θ_- curves all pass through the same values at the frequencies where $\theta_- = \frac{1}{2}\pi$ and $\frac{1}{4}\pi$, corresponding to $g = 0$ and ∞.

The corresponding transmission coefficients are shown in figure 2*b*, *d*, *f*. The general discussion of §3 leaves some doubt as to exactly what will be seen in this case. The symmetric and antisymmetric resonances have the same modal density, so it is conceivable that the effects of θ_+ and θ_- would cancel, and that there would be no frequencies of perfect transmission or perfect reflection. At the other extreme, when the constraint resonances are well separated we might expect to find one perfect reflection frequency and one perfect transmission frequency per mode. What is revealed by the calculation is intermediate between these two. For all cases of coupling strength it turns out that there is a perfect-transmission frequency for every constraint mode, but that perfect reflection occurs only every two modes. It occurs at the clamped-free half-beam frequencies, where Y_r and Y_t both vanish: these are frequencies of perfect reflection whatever the ratio of stiffness to density for the attached beam. But the fact that the θ_+ and θ_- curves just touch at those frequencies in figure 2 is specific to the case where the attached beam has the same ratio as the infinite one: for differing ratios the curves will cross, producing two frequencies of perfect reflection per pair of constraint modes.

(*b*) *Cylindrical shell with a plane baffle*

A more complicated example of the behaviour discussed in §3, and one of direct engineering significance, is a thin, circular cylindrical shell with a thin panel bridging the interior in a plane perpendicular to the cylinder generators. It is representative of, for example, a tank with a baffle or an airframe structure with a bulkhead. At first sight it might appear that the results of §3 do not apply to this system, because it is not one dimensional. However, the cylindrical symmetry of both wave-bearing system and constraint means that any possible motion of the system can be decomposed into waveguide modes, in which radial shell motion may be assumed to vary with azimuthal angle ϕ according to $\cos n\phi$ or $\sin n\phi$, where $n = 0, 1, 2, 3, \ldots$, labels the successive waveguide modes. We refer to n as the 'angular order' of a given waveguide mode.

For each angular order, considered separately, we have a one-dimensional problem of the kind discussed in §3. The plane baffle obviously satisfies the symmetry assumption. There are in general four wavetypes on a thin cylindrical shell, of which up to three may be propagating. However, for an angular order of 2 or greater, the two propagating wavetypes involving predominantly in-surface motion have cut-on frequencies significantly higher than that of the predominantly flexural wavetype. Thus for a range of low frequencies, the assumption of a single propagating wavetype is satisfied. There are then three evanescent wavetypes.

This is a system for which the constraint strength for symmetric and antisymmetric motion will be very different. Symmetric motion of the shell couples to the in-plane motion of the circular plate, whereas antisymmetric motion of the shell couples to flexural motion of the plate. When the plate is thin, one would anticipate that the flexural motion will present a high admittance to the shell, compared with the in-plane motion which will generally present a much lower admittance. Also, the modal densities of the symmetric and antisymmetric resonances of the constraining system will be quite different. The modal density of flexural modes in the plate will be much higher than that for the in-plane modes, provided again that the plate is thin.

Applying the result of §3 now produces a surprising prediction. Since the modal density of antisymmetric resonances is much greater than that of symmetric resonances, $\theta_+ - \theta_-$ will have a systematic trend dominated by the behaviour of θ_-, and there will inevitably be a sequence of frequencies of perfect reflection and perfect transmission. But these arise from the rather low-impedance flexural modes of the baffle, and in spite of the fact that the in-plane motion imposes a strong constraint on the cylinder at almost all frequencies. One might have expected some enhancement of transmission around flexural resonances, but that the in-plane constraint can be overcome entirely to produce perfect transmission is not perhaps immediately intuitive.

Detailed computation confirms this prediction. For this purpose, the required symmetric and antisymmetric impedances of the circular plate may be inferred from the classic works of Love (1927, §314) and Rayleigh (1877, §218, *et seq.*), while the modelling of the cylinder follows Arnold & Warburton (1949). As an example, we

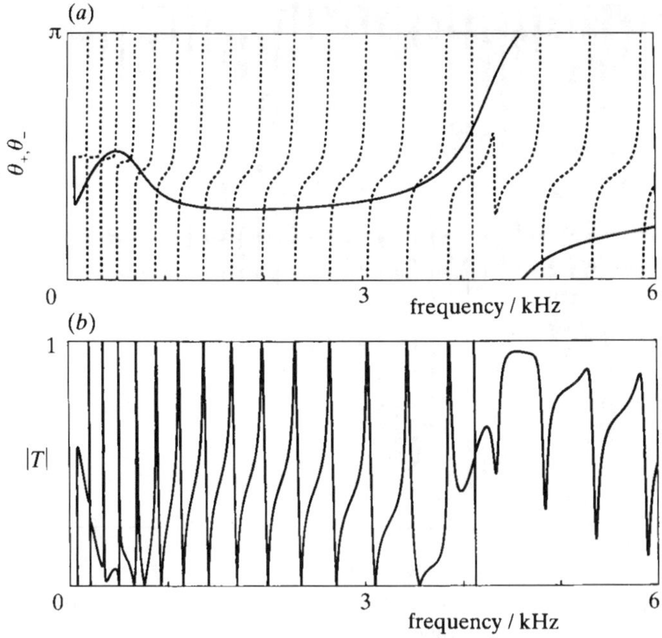

Figure 3. (*a*) Phase angles θ_+ (solid) and θ_- (dashed) and (*b*) transmission coefficient plotted against frequency, for vibration with angular order $n = 4$ on an infinite thin circular cylinder with a thin, plane circular baffle (as described in the text).

consider a steel cylinder of radius 1 m and thickness 5 mm, and a steel baffle of thickness 5 mm. The ring frequency for this cylinder lies at 867 Hz. We show first some results for a typical angular order, $n = 8$. The cut-on frequencies for the three propagating wavetypes are 78 Hz (flexural waves), and 4133 Hz and 6990 Hz (in-surface waves). There is indeed a wide range of frequencies for which only flexural waves can propagate, within which the result of §3 may be applied.

The computed transmission coefficient $T(\omega)$ is plotted in figure 3*b*. Behaviour in the expected pattern is immediately apparent, with frequencies of perfect transmission and perfect reflection occurring approximately at the modal density of flexural resonances in the baffle. The cut-on frequency for flexural waves is visible at the left of figure 3*b*. The lower cut-on for in-surface motion makes itself apparent in a more subtle way. For frequencies above this, while the transmission coefficient continues to have peaks associated with flexural resonances of the baffle, those peaks no longer reach a magnitude of unity. This is to be expected, since some of the incident energy can now be scattered into the other propagating wavetype at the baffle.

The functions θ_+ and θ_- are plotted in figure 3*a*. They present very different appearances, as would be expected from the discussion above. The in-surface motion has only a single resonance in the frequency range plotted, so that the θ_+ curve is slowly varying. The hump at low frequencies arises from the dispersion characteristics of wave propagation on the cylinder for this angular order: on the low-frequency side of the hump the cylinder is significantly stiffened by curvature effects (relative to a flat plate of the same thickness and material) (Arnold & Warburton 1949). This effect diminishes as the ring frequency is approached, and above that the behaviour is much closer to that of a flat plate. This phenomenon also produces an effect on θ_-.

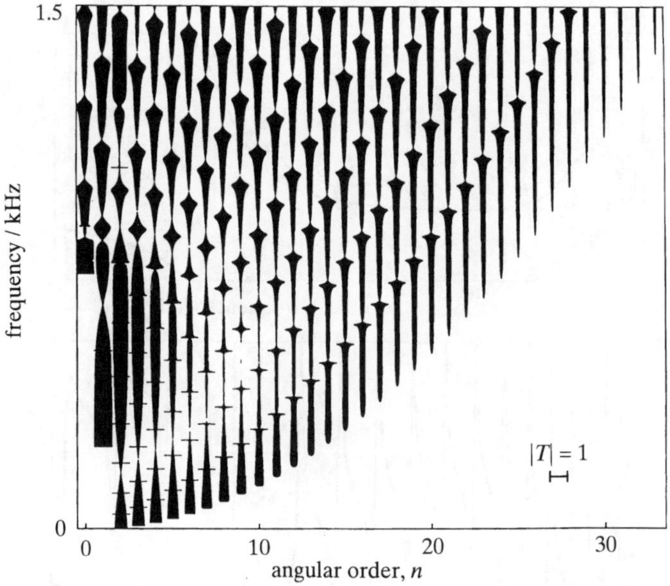

Figure 4. Transmission coefficient for the baffled cylinder problem of figure 3, plotted against frequency and angular order.

At very low frequencies, corresponding to the rise of the hump in θ_+, the peaks of perfect transmission are extremely narrow since the impedance mismatch between baffle and cylinder is quite large. Once the ring frequency is approached, the radiation damping of the baffle modes by the infinite cylinder increases and the peaks become broader.

The computation may be repeated for the other angular orders. The resulting transmission coefficients for $n = 0, 1, \ldots, 33$ are all plotted in figure 4, for frequencies up to 1500 Hz. For each value of n, the result is represented by a vertical stripe whose width is modulated in proportion to the transmission coefficient magnitude. The stripes for $n = 0$ and 1 are truncated at low frequencies, when the associated cylinder displacement becomes predominantly in-surface rather than radial. The frequencies of perfect transmission appear as the wide spots on each stripe, and the systematic variation of these frequencies with n is evident in the figure. The curves traced out by these perfect transmission frequencies can be thought of as lines along which trace wavenumber coincidence occurs between waves on the cylinder and in the baffle.

5. Implications for statistical energy analysis

(a) *The effect on variance of* SEA *estimates*

Once the transmission coefficient through a junction has been calculated, the standard SEA procedure is to turn this into a coupling loss factor by various averaging processes. These include some or all of: frequency averaging over a chosen bandwidth; ensemble averaging over systems whose properties and dimensions are drawn from a statistical population representing manufacturing tolerances; and (if the junction is extended in one or two dimensions) averaging over angles of incidence of waves on the boundary, assuming a diffuse field within the source subsystem. Frequency averaging and ensemble averaging may in fact be the same process, if an ergodic assumption is made about the effect of manufacturing differences on the subsystem frequency response functions.

For a resonant coupling of the kind investigated here, the first issue to consider is the bandwidth for frequency averaging, compared with the expected spacing of frequencies of perfect transmission. The usual SEA philosophy is to make sure that the averaging bandwidth is wide enough to contain at least a few modes of each subsystem, because the essence of SEA is to lose the unwanted detail of individual modal responses. One might therefore suppose that the same argument should be applied to the 'modes', that is the frequencies of perfect transmission, of any junctions in the system. Averaging over a very narrow band will give a coupling loss factor which has significant frequency variation (Woodhouse 1981), although if an ensemble average is also carried out this variation may be smoothed out.

But if the junction transmission coefficient is averaged over a sufficiently wide band that a smoothly-varying answer is obtained, then a trap has been concealed which may have a profound effect on the overall accuracy of the SEA model. The effect is most striking for a junction which behaves like figure 2*f*, with a transmission coefficient which is generally low except for narrow peaks at which it reaches unity. The calculated coupling loss factor will then be quite small, leading one to expect a situation with 'weak coupling', for which SEA is always said to work well (Lyon 1975; Hodges & Woodhouse 1986). But the energy transmission by such a junction is very selective: the source subsystem may have a uniform distribution of vibrational energy in frequency, but the transmitted vibration will be strongly concentrated in the narrow bands where the transmission is high.

If only these two subsystems are involved, this filtering may not matter very much. A correct SEA model should predict the mean-square level in the second subsystem accurately. The difficulty arises if this second subsystem is coupled to a third by another junction with resonant properties. Then one of two things can happen, and neither of them is handled very well by conventional SEA. If the second junction is the same as the first, the energy incident on it is 'pre-filtered' to fit its pattern of strong transmission. So one can expect a much higher proportion of the energy to be transmitted across the second junction than across the first. This leads to a phenomenon of 'diminishing returns' if the system geometry has a chain of similar couplings. Explicit analysis of this phenomenon, based on two quite different approaches, has been given by Heron (this book) and Langley & Bercin (this book).

If the second junction is different from the first, a different problem arises. Typically, the very small number of perfect-transmission peaks within a given frequency band will not line up in the two junctions, so that the field which has been filtered by the first junction will transmit less well through the second than would have been the case for a spectrally-white incident field. The normal, linearly-averaged SEA coupling loss factor is not a very good measure of this energy transmission. If the conventional coupling loss factors are used, this phenomenon will produce an increase in the variance around the SEA mean prediction, when different members of the statistical ensemble are tested. It is possible that this variance could be reduced by a different choice of averaging procedure for the coupling loss factors (Hodges & Woodhouse 1986).

Similar considerations apply to a junction which is extended in one or two dimensions. For example, the data of figure 4 can be used to calculate a coupling loss factor. As well as a frequency average, a suitable weighted sum of the results for different angular orders must be taken, to represent a diffuse field incident on the

Junction (Lyon 1975 §3.3; Langley 1994). Perfect transmission occurs at different frequencies for the different values of n, and one does not require such a broad averaging bandwidth to achieve a smoothly varying random-incidence transmission coefficient, from which a coupling loss factor may be obtained. Filtering effects are produced by this junction too, not only filtering in frequency but also in circumferential wavenumber. This can produce similar effects to those noted for the beam example, for example diminishing returns from successive transmission through identical junctions (Langley & Bercin, this book).

(b) *When is a junction a subsystem?*

There is a second general issue raised by junctions with internal degrees of freedom: under what circumstances does it become necessary to treat the coupling as an energy-storing subsystem in its own right? Of course, if one needs to allow for damping in the junction structure, or excitation on it, then the SEA formalism requires that it appears as a subsystem. Problems associated with low modal density may then arise, but those are familiar and are not the subject of this investigation. We consider here the question of accuracy of modelling: the example from §4*a* will be analysed as a two-subsystem problem with complicated coupling and as a three-subsystem problem with simpler coupling, and the results compared.

Consider first the general problem of two joined systems, with energies E_1 and E_2 and modal densities n_1 and n_2, having a third system attached symmetrically at the junction point. This third system is assumed to have no damping or external drive, and to have energy E_3 and modal density n_3. Suppose that the subsystems 1 and 2 are locally physically similar (for example, two sections of beam with the same cross-section but perhaps different lengths). Using the normal SEA formalism (Lyon 1975 §3.2), a requirement of power balance on the third subsystem gives

$$E_3/n_3 = \tfrac{1}{2}[E_1/n_1 + E_2/n_2] \tag{5.1}$$

(where advantage has been taken of the symmetry of coupling $1 \leftrightarrow 3$ and $2 \leftrightarrow 3$). In terms of the thermal analogy of SEA, the attached system adjusts to the mean 'temperature' of the other two subsystems. This equation may be used to eliminate E_3 and n_3 from the other two power-balance equations, and the result is an 'equivalent two-subsystem model' derived from the three-subsystem model. If the coupling factors (i.e. the products of coupling loss factor and modal density which satisfy reciprocity) in the three-subsystem model are ϵ_α (between systems 1 and 2) and ϵ_β (between systems 1 or 2 and system 3), then the equivalent coupling factor between systems 1 and 2 in the two-subsystem model turns out to be $(\epsilon_\alpha + \frac{1}{2}\epsilon_\beta)$.

To apply this to the example of §4*a*, we first calculate the transmission coefficients for the problem in which the two halves of the attached cantilever beam are regarded as semi-infinite. These can be used to calculate the coupling factors ϵ_α and ϵ_β by the usual SEA wave-method approach. But for the present purpose it is easier to work in reverse and deduce an equivalent transmission coefficient from the argument given above, which can be compared directly with a suitably averaged value of the transmission coefficients plotted in figure 2.

The calculation of the transmission coefficients at a junction of four semi-infinite beams is quite straightforward. The general expressions for the answers are quite lengthy, but when the assumption is made, as before, that the 'attached' beams have bending stiffness λB and line density λm (compared to the corresponding properties B and m of the original semi-infinite beams) they reduce to very simple expressions.

The (energy) transmission coefficient between the two sections of original beam is $\tau_1 = 1/(1+\lambda)^2$, and that from a section of original beam to a section of 'attached' beam is $\tau_2 = \lambda/(1+\lambda)^2$. Noting that the attached subsystem is composed of both (identical) sections of attached beam, the required equivalent transmission coefficient, based on $(\epsilon_\alpha + \frac{1}{2}\epsilon_\beta)$ from above, is then

$$\tau_1 + \tau_2 = 1/(1+\lambda). \tag{5.2}$$

So for the case of figure 2*f* with $\lambda = 10$, the equivalent transmission coefficient is 1/11, −10.41 dB.

When a frequency average is performed on the results plotted in figure 2*f*, choosing one 'cycle' of the obvious pattern as the bandwidth, the result rapidly converges to 0.9836, −10.07 dB. (At very low frequencies it is slightly different, largely because of the influence of the evanescent fields.) So for this example, the two methods agree with some accuracy. Numerical experiments reveal that this agreement persists over a very wide range of assumed properties of the attached beam. So at least for this idealized problem, it seems that one can treat the attached beam as a complicated coupling or as a third subsystem, and obtain essentially the same SEA model by either route. It would be interesting to test this conclusion on other combinations of wavebearing system and junction structure.

6. Conclusions

Resonant couplings of the kind examined here produce strong frequency dependence of the transmission coefficient, and this can have significant implications for the result and interpretation of vibration analysis, whether deterministic or statistical, of a system which includes the junction. Frequencies of perfect transmission and of perfect reflection are likely to occur, and these are a universal feature of junctions of the kind considered here. Modifications to the detailed design may move them around, but will not eliminate them unless they break the assumed symmetry. (An example of the possible usefulness of broken symmetry is the improved transmission loss when two panes of glass of different thickness are used in double glazing, compared with two identical panes.) Insight can be gained by considering the physical nature of the vibration modes of the junction. Symmetric and antisymmetric modes should be considered separately, and their modal densities and strength of coupling to the rest of the system examined.

References

Arnold, R. N. & Warburton, G. B. 1949 Flexural vibration of the walls of thin cylindrical shells having freely supported ends. *Proc. R. Soc. Lond.* A **197**, 238–256.

Cremer, L., Heckl, M. & Ungar, E. E. 1973 *Structure-borne sound*. Berlin: Springer.

Hodges, C. H. & Woodhouse, J. 1986 Theories of noise and vibration transmission in complex structures. *Rep. Prog. Phys.* **49**, 107–170.

Langley, R. S. 1994 Elastic wave transmission coefficients and coupling loss factors for structural junctions between curved panels. *J. Sound Vib.* **169**, 297–317.

Love, A. E. H. 1927 *A treatise on the mathematical theory of elasticity*. (Reprinted by Dover (New York) 1944.)

Lyon, R. H. 1975 *Statistical energy analysis of dynamical systems*. Cambridge, Massachusetts: MIT Press.

Rayleigh, Lord 1877 *The theory of sound.* (Reprinted by Dover (New York) 1945.)

Skudrzyk, E. 1980 The mean-value method of predicting the dynamic response of complex vibrators. *J. Acoust. Soc. Am.* **67**, 1105–1135.

Woodhouse, J. 1981 An approach to the theoretical background of statistical energy analysis applied to structural vibration. *J. Acoust. Soc. Am.* **69**, 1695–1709.

Statistics of energy flows in spring-coupled one-dimensional subsystems

By C. S. Manohar and A. J. Keane
University of Oxford, Department of Engineering Science, Parks Road, Oxford OX1 3PJ, U.K.

This paper considers the problem of determining the statistical fluctuations occurring in the vibrational energy flow characteristics of a system of two multimodal, random, one-dimensional subsystems coupled through a spring and subject to single frequency forcing. The subsystems are modelled either as transversely vibrating Euler–Bernoulli beams or as axially vibrating rods. The masses of the subsystems are modelled as random variables. The calculations of energy flows are based on an exact formulation which uses the Green functions of the uncoupled subsystems, which, in turn, are expressed as summations over the uncoupled modes. Factors influencing the number of modes contributing to the response statistics at any specified driving frequency are investigated. A criterion for identifying the driving frequency beyond which the mean power spectra become smooth is proposed. Empirical procedures are developed to predict the 5% and 95% probability points given knowledge of the first two moments of the response. The work reported here forms part of a long term study into the reliability of statistical energy analysis (SEA) methods.

1. Introduction

The process of averaging in statistical energy analysis (SEA) is carried out for two main reasons: first, it accounts for the random nature of the forces exerted on most structures, thereby simplifying measures of response; and second, it caters for the stochastic modelling of the system which is adopted to allow for the sensitivity of high-frequency responses to minor changes in physical and modelling parameters. It must also be noted that the primary response variables of interest in SEA are the steady state average total energies stored in the subsystems. These quantities are obtained as integrals over the extent of the subsystems and also over driving frequency bands, which imply a further combination of both spatial and frequency averaging. The results obtained are clearly dependent on details of the averaging processes such as the frequency bandwidths, quantities treated as random and the probability distribution functions assumed for these random quantities. Each form of averaging is accompanied by a reduction in the resolution of the response with respect to amplitude, time, space or frequency parameters. This is, of course, consistent with the primary aim of SEA modelling, which is to produce simplified models of system behaviour which describe gross properties of system responses. However, in order that the average results can be interpreted properly, especially with respect to observations made on a single realization of a system or an excitation over a limited time or frequency interval, it is essential that each process of averaging be accompanied by associated estimates of the measures of dispersion. While it is fairly straightforward to analyse the dispersion associated with averaging across an ensemble of time histories using standard random vibration theory (see Lin 1967), the study of other forms of averaging is considerably more complicated. This difficulty constitutes a major shortcoming in the application of SEA procedures to practical problems and has received relatively little attention in the literature. Work has so far been carried out in this direction by Lyon & Eichler (1964), Lyon (1969), Davies & Wahab (1981), Davies & Khandoker (1982), Fahy & Mohammed (1992) and Keane & Manohar (1993). The studies conducted by Skudrzyk (1968, 1980, 1987) on the bounds of system transfer functions can also be cited in this context. A discussion on related issues can also be found in the works of Scharton & Lyon (1968), Hodges & Woodhouse (1986), Heron (1990) and Craik (1991).

The present study considers this problem by investigating the stochastic variability of energy flows in a system of coupled beams or rods. The statistics of the dissipated powers in the individual subsystems are investigated as functions of driving frequency, for the cases of point harmonic and rain-on-the-roof type distributed excitations (i.e. for single frequency driving). The bands enclosing the 5% and 95% probability points are shown to display different types of behaviours, namely, oscillatory, convergent, divergent or stationary, depending on the choice of subsystem type (i.e. beams or rods), damping models, type of excitation and details of the stochastic model used for the system. A non-dimensional parameter related to the variability in the subsystem natural frequencies is introduced which is shown to be useful in characterizing the frequency beyond which the resonant behaviour of individual modes no longer dominates the response statistics. Clearly, this frequency represents a cutoff point below which simplified theories like SEA cannot be guaranteed to work well. The problem of estimating confidence bands empirically using knowledge of the first two moments is also investigated. This study shows that the energy flow statistics can be described reasonably well using either gamma or lognormal probability distribution functions.

2. The two-subsystem model

The system under consideration consists of a pair of transversely vibrating beams or axially vibrating rods, which are mutually coupled through a spring, the system configuration being illustrated in figure 1. The subsystems are assumed to have random material and/or geometrical properties and are taken to be viscously damped. No restrictions are placed either on the magnitude of damping or the strength of the coupling spring. The external excitations acting on the system are modelled either as point harmonic forces or as a rain-on-the-roof type distributed forcing. The aim of the present study is to examine the probabilistic nature of the energy flow characteristics in the system arising out of random fluctuations in the system properties. The deterministic aspects of energy flow characteristics in this type of system have been studied by Davies (1973) and Keane (1992). The expressions for the receptance functions and the input, dissipated and coupling powers are readily available in these references and hence are reproduced here without detailed derivation. Thus, when two subsystems are coupled at $x_i = a_i$ and excited by point forces $F_i(t)$ acting at $x_i = b_i$, $i = 1, 2$, the input power and coupling power receptances for the first subsystem are given respectively by

$$H_{\mathrm{in}\,1}(\omega) = (\omega^2/m_1) \sum_{i=1}^{\infty} c_{1i}(\psi_i(b_1)/|\phi_i|^2) + (\omega k_c/m_1^2)\,\mathrm{Im}\left\{\frac{1}{\Delta}\left[\sum_{i=1}^{\infty} \psi_i(b_1)\,\psi_i(a_1)/\phi_i\right]^2\right\} \quad (1)$$

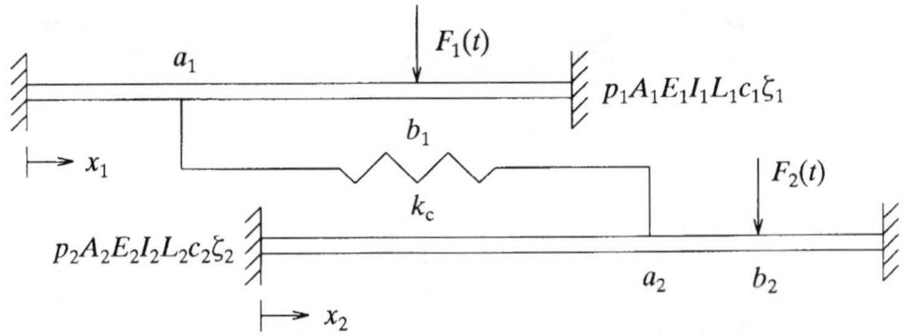

Figure 1. Two spring coupled one-dimensional subsystems.

and
$$H_{12}(\omega) = \frac{\omega^2 k_c^2}{m_1^2 m_2 |\Delta|^2} \sum_{r=1}^{\infty} c_{2r}(\psi_r^2(a_2)/|\phi_r|^2) \left| \sum_{i=1}^{\infty} (\psi_i(b_1)\,\psi_i(a_1)/\phi_i) \right|^2. \quad (2)$$

Similarly, when the system is excited by distributed forces $F_i(x_i, t), i = 1, 2$, of the rain-on-the-roof type, the above receptances are given by

$$H_{\text{in}\,1}(\omega) = (\omega^2/m_1) \sum_{i=1}^{\infty} c_{1i}(1/|\phi_i|^2) + (\omega k_c/m_1^2)\,\text{Im}\left\{ \sum_{i=1}^{\infty} (\psi_i^2(a_1)/\phi_i^2\,\Delta) \right\} \quad (3)$$

and
$$H_{12}(\omega) = \frac{\omega^2 k_c^2}{m_1^2 m_2 |\Delta|^2} \sum_{r=1}^{\infty} c_{2r}(\psi_r(a_2)/|\phi_r|^2) \sum_{i=1}^{\infty} (\psi_i^2(a_1)/|\phi_i|^2). \quad (4)$$

In these equations the summations over the indices i and r respectively denote summations over the modes of the first and second subsystems. The quantities ω_i and ψ_i are the natural frequencies and mode shapes with the quantities ϕ_i and Δ given by

$$\phi_i = \omega_i^2 - \omega^2 + ic_{1i}\,\omega \quad \text{and} \quad \Delta = 1 + (k_c/m_1) \sum_{i=1}^{\infty} (\psi_i^2(a_1)/\phi_i) + (k_c/m_2) \sum_{r=1}^{\infty} (\psi_r^2(a_2)/\phi_r). \quad (5, 6)$$

The mode shapes ψ satisfy the orthogonality conditions given by

$$\int \psi_i(x_1)\,\psi_j(x_1)\,\rho_1(x_1)\,\mathrm{d}x_1 = m_1\,\delta_{ij}. \quad (7)$$

Here δ_{ij} denotes the Kronecker delta function. The quantities m_i and ρ_i denote the total mass and mass per unit length of the ith subsystem. c_{ij} is the damping coefficient of the ith subsystem in the jth mode; k_c is the coupling spring constant. The forces F_1 and F_2 are assumed to be ergodic and statistically independent. The spectra of the input, coupling and dissipated powers can be related using the above receptance functions as follows:

$$\Pi_{\text{in}\,1}(\omega) = H_{\text{in}\,1}(\omega)\,S_{F_1}(\omega), \quad \Pi_{12}(\omega) = H_{12}(\omega)\,S_{F_1}(\omega) - H_{21}(\omega)\,S_{F_2}(\omega) \quad (8, 9)$$

and
$$\Pi_{\text{diss}\,1}(\omega) = \Pi_{\text{in}\,1}(\omega) - \Pi_{12}(\omega). \quad (10)$$

Here $S_{F_i}(\omega)$ is the power auto-spectral density function of $F_i(t)$.

When the mass and/or stiffness properties of the individual subsystems are modelled as random quantities, the natural frequencies and mode shapes become random in nature. The functions described above, in turn, become random processes. The aim of the present investigation is to obtain probabilistic descriptions of the different power functions given in equations (8)–(10) as functions of the probabilistic descriptions of the masses of the individual rods. Estimates for the probability distribution functions (PDFS) of these quantities can currently only be obtained using Monte Carlo simulation techniques. For this purpose, an ensemble of realizations of coupled subsystems are computationally simulated as per the stochastic model adopted. For every realization of the pairs of subsystems, the natural frequencies and mode shapes are calculated for the individual subsystems and this information incorporated into equations (1)–(4) to generate the ensemble of receptance functions. This ensemble is further processes to obtain the desired PDFS.

For simple theories like SEA to be useful, it is clearly necessary that the mean spectra of the different power functions become stationary with increases in the frequency of interest and moreover, the 5% and 95% probability points should preferably converge towards the mean. For any given problem it is not obvious at the outset whether such behaviour occurs and this study identifies some of the more influential factors.

3. Modal overlap and statistical overlap factors

The receptance functions and power spectra have been expressed in equations (1)–(10) in terms of summations over the modes of the uncoupled subsystems. At any specified frequency, the number of modes making significant contributions to the response is clearly dependent on the bandwidth of the nearby modes and the modal frequency spacing. The ratio of these two quantities has been defined in the literature as the modal overlap factor, see Lyon (1975). The modal spacing is governed by the type of subsystem considered; for example, it remains constant with frequency for axially vibrating rods and increases linearly for transversely vibrating Euler–Bernoulli beams. The modal bandwidth, on the other hand, is a function of the damping model adopted for the subsystem. In SEA studies the damping is normally taken to be viscous and proportional. Within the framework of this assumption several alternatives are possible which can dramatically alter the behaviour of the receptances and power spectra. Thus, if the damping force per unit length is expressed as $r(x)\,\dot{y}(x,t)$, the damping coefficient $r(x)$ can be taken to be proportional to local mass, stiffness or a linear combination of mass and stiffness. Depending on the model used, the modal damping factor, ζ_n, can either fall or rise with the mode count n. Another alternative which is commonly employed, is to take ζ_n to be a constant for all modes. In this study the damping is taken to be either mass proportional, leading to modal damping factors falling with frequency and constant modal bandwidths, B_n, or to be represented by constant damping factors. These two damping models are summarized in table 1 where the relevant expressions for simply supported beams and fixed–fixed rods are given. It may be observed from the table that for constant B_n, the model overlap factor remains fixed for rods while it varies as n^{-1} for beams; conversely for constant ζ_n, the overlap factor varies as n for both beams and rods.

It should be noted here that the strength of random fluctuation assumed for the subsystem mass parameters has a significant effect on the number of modes contributing to the statistics of the spectra at any specified frequency. In order to see this, consider subsystems in which the mass per unit length ρ is uniform and modelled as

$$\rho = \rho_0(1+\epsilon U), \tag{11}$$

where U is a gaussian random variable with zero mean and unit standard deviation and ϵ controls the degree of randomness. Figure 2 shows the resulting probability

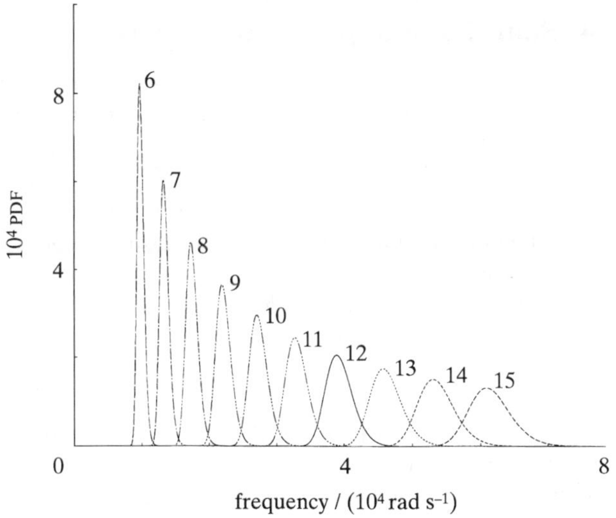

Figure 2. Probability density functions for natural frequencies of beam 1; $\epsilon = 0.10$. Values of n as indicated on figure.

Table 1. *Modal properties of rods and beams*

quantity	B_n	ζ_n	beam	rod
natural frequency	—	—	$n^2\pi^2/L^2\sqrt{(EI/\rho)}$	$n\pi/L\sqrt{(AE/\rho)}$
modal spacing	—	—	$\pi^2/L^2(1+2n)\sqrt{(EI/\rho)}$	$\pi/L\sqrt{(AE/\rho)}$
modal overlap factor, M	c	$c/2\omega_n$	$cL^2/\pi^2(1+2n)\langle\sqrt{(EI/\rho)}\rangle$	$cL/\pi\langle\sqrt{(AE/\rho)}\rangle$
	$2\zeta\omega_n$	ζ	$2\zeta n^2/1+2n$	$2\zeta n$
statistical overlap factor, S_n	—	—	$2\beta n^2/1+2n$	$2\alpha n$

$$M = \frac{\text{mean modal bandwidth}}{\text{mean modal spacing}}, \quad S_n = \frac{2\sigma_n}{\langle\omega_{n+1}-\omega_n\rangle},$$

$$\alpha = \sqrt{[\langle AE/\rho\rangle - \langle\sqrt{(AE/\rho)}\rangle^2]}/\langle\sqrt{(AE/\rho)}\rangle, \quad \beta = \sqrt{[\langle EI/\rho\rangle - \langle\sqrt{(EI/\rho)}\rangle^2]}/\langle\sqrt{(EI/\rho)}\rangle.$$

density functions of a set of ten consecutive natural frequencies for a simply supported beam (a similar plot is obtained for the case of a fixed–fixed rod, except that the frequency spacing is then constant). It can be seen from this figure that the probability density functions 'overlap' increasingly for higher frequencies. This overlap will clearly be greater for larger ϵ. This implies that the randomness parameter ϵ plays a role similar to that of the damping bandwidth in the sense that, at any specified frequency, an increase in ϵ results in a larger number of modes contributing to the *statistics* of the response. This behaviour is not reflected clearly in the definition of the model overlap factor although it is weakly dependent on ϵ. Consequently, it is useful to introduce a statistical overlap factor defined by

$$S_n = 2\sigma_n/\langle\omega_{n+1}-\omega_n\rangle, \tag{12}$$

where σ_n is the standard deviation of the nth natural frequency. From table 1 it can be seen that this overlap factor increases as n for both rods and beams. As will be seen in the later sections, S_n can be shown to be related to the frequency beyond which the oscillations in the *statistics* of the power spectra die out. It may be noted that this quantity is closely related to the 'spacing signal to noise ratio' discussed by Soong & Bogdanoff (1963) in the context of simple oscillator chains, where departures from the expected values of natural frequencies were studied.

4. Statistics of response power spectra

The factors likely to influence the number of modes contributing to responses at any specified frequency have been discussed in the previous section. The results of a parametric survey on the statistics of the response power spectra are presented in this section, highlighting the role played by these factors. The study of such response statistics is currently not feasible using analytical procedures and here numerical simulation methods have been used. There are three main questions of interest which this survey attempts to address.

1. Under what conditions do the ensemble mean square responses become stationary with respect to the driving frequency?

2. If they do, what is the frequency beyond which this behaviour can be expected to occur?

3. Do the contours of 5% and 95% probability points converge onto the mean, become constant or diverge, with increases in driving frequency?

Various factors pertaining to subsystem type, damping, stochastic modelling, excitation and strength of coupling can be expected to influence the answers to these questions. Of these, the consequences of the stochastic model employed for the subsystems are perhaps the most difficult to predict. In an earlier study we considered the energy flow characteristics in a system of two coupled, axially vibrating stochastic rods and examined different types of random system models (Keane & Manohar 1993). For most, but not all types of randomness, the transfer functions were shown to become stationary above certain values of driving frequency; equation (11) being of this type. The present study is not focused on the details of different randomization schemes and so this model is adopted throughout. None the less, failure to exhibit stationarity should not be discounted when attempting to apply SEA methods.

The remaining factors of interest have been studied by carrying out a full survey based on the system of two coupled subsystems already illustrated in figure 1; table 2 details the subsystem properties used. The effects of varying various parameters in the problem were considered as follows: subsystem type (Euler–Bernoulli beams or axially vibrating rods), damping models (constant modal bandwidth or constant modal damping coefficient), magnitude of damping (see table 3), levels of system randomness ($\epsilon = 0.01$, 0.05 or 0.10) and type of excitation (point harmonic or distributed rain-on-the-roof excitation). It may be observed from table 3 that the parameter values for the two damping models have been chosen so that the 10th mode of rod vibration and 4th mode of beam vibration both have identical damping factors and modal bandwidths for subsystem 1; these modes lying 25% of the way through the frequencies considered here. Throughout this survey the energies dissipated in the two subsystems were investigated for the case when the first subsystem was excited by external forces with the second subsystem driven only indirectly through the coupling spring. Consequently, the power dissipated in the undriven subsystem is proportional to the cross power receptance function while that dissipated in the driven subsystem reflects both the input and cross power receptances, see again equations (8)–(10). The coupling spring constants were chosen as $k_c = 0.9 \times 10^9\ \mathrm{N\,m^{-1}}$ for the rod systems and $k_c = 0.1 \times 10^9\ \mathrm{N\,m^{-1}}$ for the beam systems to ensure that the subsystems were well coupled to each other (i.e. so that the behaviour of the undriven subsystem significantly affects that of the driven one (see Keane 1992)). An excitation frequency range from 0 to 50000 rad $\mathrm{s^{-1}}$ was

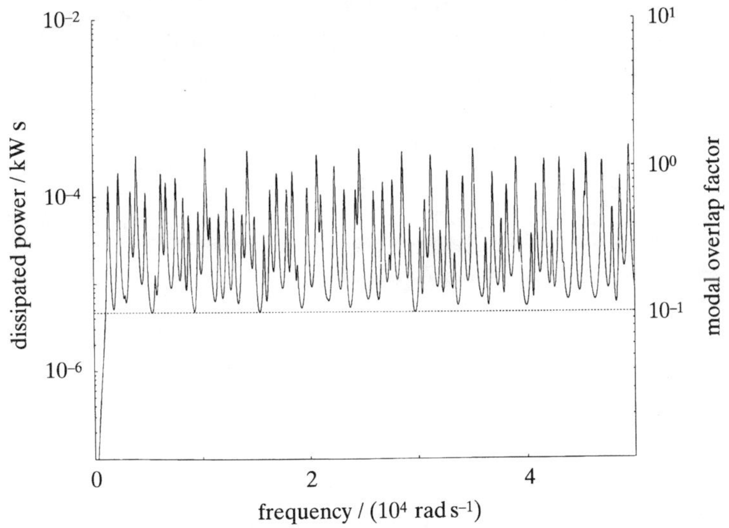

Figure 3. Dissipated power spectral density function for subsystem 1 of the rod system; rain-on-the-roof excitation; $c = 130\ \mathrm{s}^{-1}$. ——, power; · · · · · ·, modal overlap factor.

Table 2. *Subsystem physical properties*

subsystem	length/m	rigidity	mass/unit length (kg m^{-1})	drive point, a/m	coupling point, b/m
rod 1	5.0	17.85 MN	4.156	1.15	3.15
rod 2	4.5	17.85 MN	4.156	—	2.115
beam 1	1.0	1536.5 N m^2	2.0141	0.23	0.63
beam 2	0.9	1536.5 N m^2	2.0141	—	0.423

Table 3. *Subsystem damping properties*
(Rod 2 and beam 2 have equivalent values of B_n and ζ_n.)

system	bandwidth	damping coefficient
rod 1	$B_1 = 13.0$, $B_{10} = 130.2$, $B_{40} = 520.8$	$\zeta_n = 0.005$
	$B_1 = 78.1$, $B_{10} = 781.2$, $B_{40} = 3124.8$	$\zeta_n = 0.03$
	$B_n = 130.0$	$\zeta_1 = 0.05$, $\zeta_{10} = 0.005$, $\zeta_{40} = 0.0012$
	$B_n = 780.0$	$\zeta_1 = 0.31$, $\zeta_{10} = 0.03$, $\zeta_{40} = 0.007$
beam 1	$B_1 = 2.7$, $B_4 = 43.6$, $B_{14} = 534.3$	$\zeta_n = 0.005$
	$B_1 = 16.4$, $B_4 = 261.9$, $B_{14} = 3205.7$	$\zeta_n = 0.005$
	$B_n = 44.0$	$\zeta_1 = 0.08$, $\zeta_4 = 0.005$, $\zeta_{14} = 0.0004$
	$B_n = 262.0$	$\zeta_1 = 0.48$, $\zeta_4 = 0.03$, $\zeta_{14} = 0.0024$

considered and calculations carried out at 200 uniformly spaced frequencies within this range. Modal summation bandwidths of 16000 rad s^{-1} for the rod systems and 100000 rad s^{-1} for the beam systems were used throughout.

The effects of changing the damping model on the *deterministic* spectra of the dissipated power (i.e. for $\epsilon = 0$) are examined in figures 3 and 4 for the case of rod systems. The extrema observed in the spectra given in these figures are governed not only by the subsystem natural frequencies but also by the mode shapes at the points of driving and coupling. Notice that the spectra of the driven subsystems given here

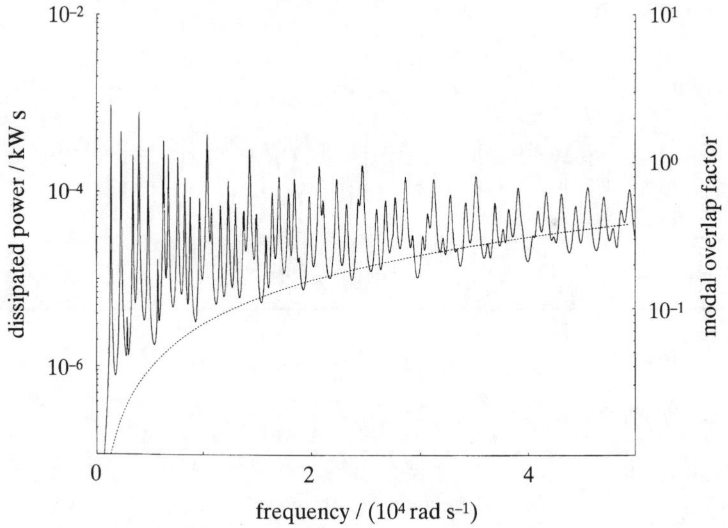

Figure 4. Dissipated power spectral density function for subsystem 1 of the rod system; rain-on-the-roof excitation; $\zeta = 0.005$; key as in figure 3.

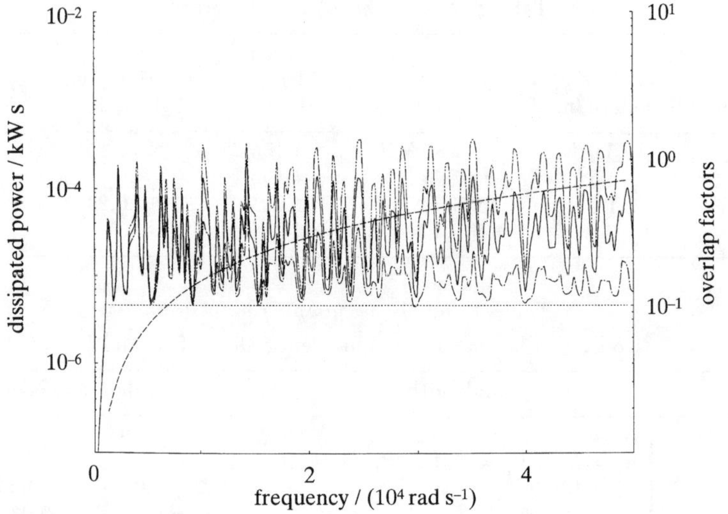

Figure 5. Statistics of the dissipated power spectral density function for subsystem 1 of the rod system; rain-on-the-roof excitation; $\epsilon = 0.01$; $c = 130\ \mathrm{s}^{-1}$; ——, mean; —·—·—, 5% probability point; —··—··—, 95% probability point; ----------------, model overlap factor; — — — —, statistical overlap factor.

are influenced by the natural frequencies of the non-driven subsystems and also by the mode shapes at the points of coupling, these effects being indicative of the strong coupling between the two subsystems. With the constant bandwidth damping model, the modal overlap factor remains constant with respect to frequency and, accordingly, the range of the spectra also remains constant (i.e. the maximum minus the minimum values). Beam systems with constant damping bandwidths behave similarly but the increasing modal spacing reduces the modal overlap factor as the driving frequency rises and, in consequence, the range widens with these increases. For the constant damping ratio model, the modal overlap factor increases with

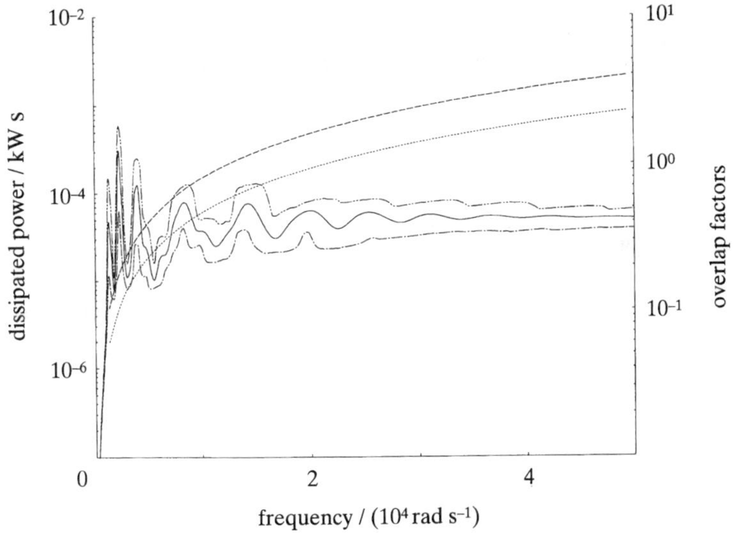

Figure 6. Statistics of the dissipated power spectral density function for subsystem 1 of the rod system; point harmonic excitation; $\epsilon = 0.10$; $\zeta = 0.03$; key as in figure 5.

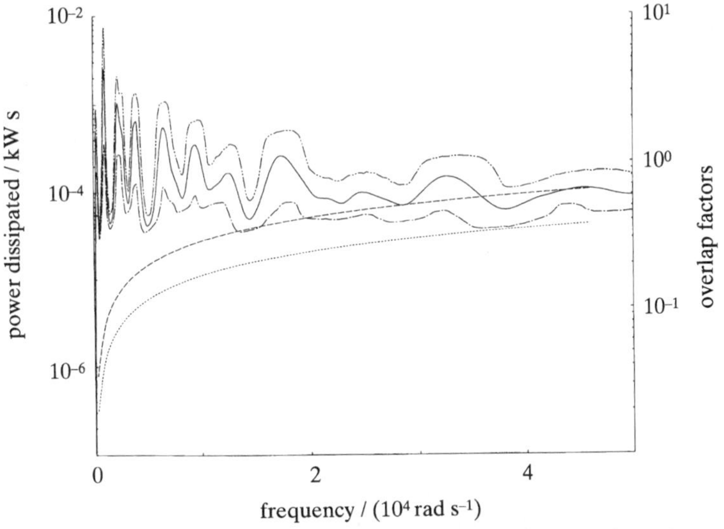

Figure 7. Statistics of the dissipated power spectral density function for subsystem 1 of the beam system; rain-on-the-roof excitation; $\epsilon = 0.10$; $\zeta = 0.03$; key as in figure 5.

frequency for both beams and rods. Consequently, the spectra in such cases tend to become smooth at higher frequencies. Clearly, for all cases, the range is found to reduce with increases in the value of modal overlap factor.

The *statistics* of the response spectra have been estimated using Monte Carlo simulation procedures with 2500 samples and results obtained for the parameter variations mentioned above. A subset of these results are presented in figures 5–12 which display the principal features of the spectral statistics with respect to the different parameters explored. In all cases only the power dissipated in the driven subsystem is plotted: as has already been noted this quantity depends on both the input of energy and its transmission to the undriven subsystem; for the strong

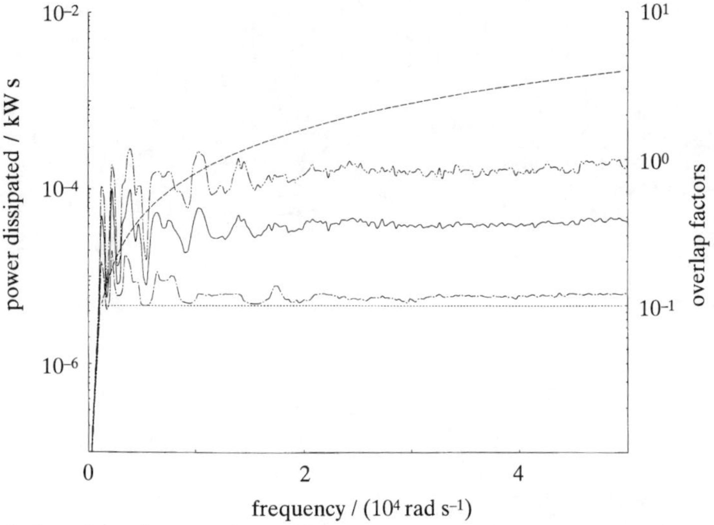

Figure 8. Statistics of the dissipated power spectral density function for subsystem 1 of the rod system; rain-on-the-roof excitation; $\epsilon = 0.10$; $c = 130\ \text{s}^{-1}$; key as in figure 5.

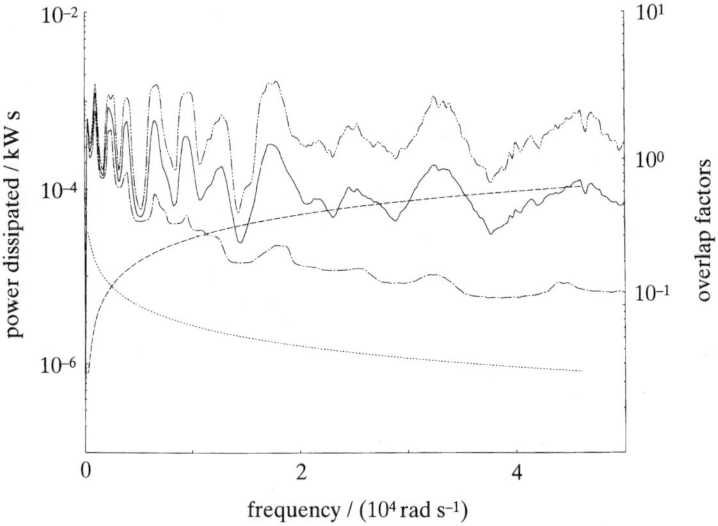

Figure 9. Statistics of the dissipated power spectral density function for subsystem 1 of the beam system; rain-on-the-roof excitation; $\epsilon = 0.10$; $c = 262\ \text{s}^{-1}$; key as in figure 5.

coupling strengths used here these quantities are of roughly equal magnitude and their statistical properties are found to be similar.

1. For low values of ϵ, the trends of the statistics closely follow the corresponding deterministic cases (cf. figures 3 and 5), which is consistent with the fact that, as $\epsilon \to 0$, the statistical solution converges to the corresponding deterministic result. The variability in the response is seen from figure 5 to be greater at higher driving frequencies and this is consistent with the greater variability found in the higher natural frequencies.

2. The most dramatic qualitative change in the behaviour of the 5% and 95% probability points is caused by the choice of damping model. For the constant ζ_n model, the contours of the probability points tend to converge onto the mean for

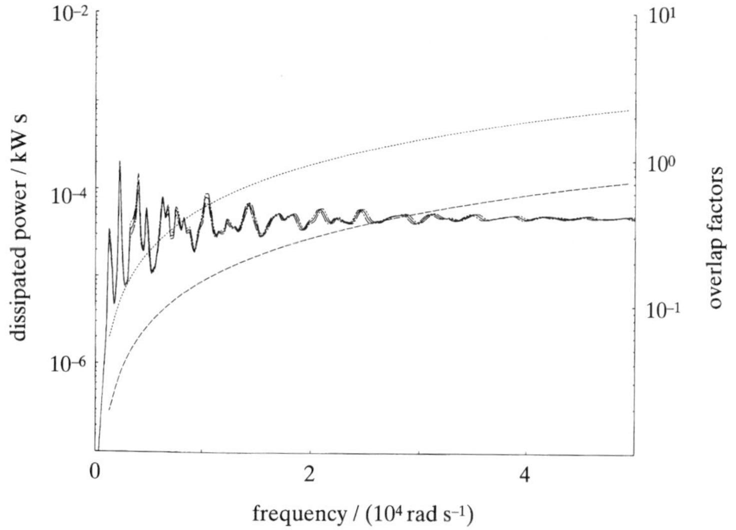

Figure 10. Statistics of the dissipated power spectral density function for subsystem 1 of the rod system; rain-on-the-roof excitation; $\epsilon = 0.01$; $\zeta = 0.03$; key as in figure 5.

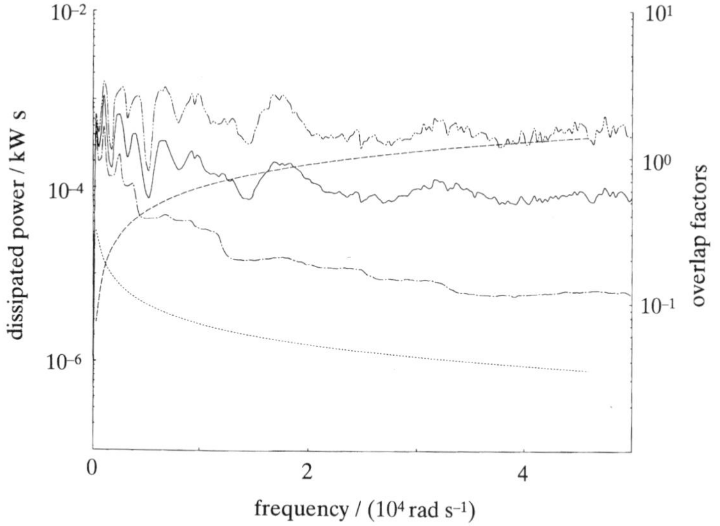

Figure 11. Statistics of the dissipated power spectral density function for subsystem 1 of the beam system; rain-on-the-roof excitation; $\epsilon = 0.20$; $c = 262\ \mathrm{s}^{-1}$; key as in figure 5.

both the cases of rod and beam systems (figures 6 and 7). This behaviour is associated with increases in the modal overlap factor arising from increases in bandwidth with driving frequency, with the convergence being faster for systems with greater modal overlap factor. On the other hand, for constant B_n, the probability points tend to become constant for rod systems (figure 8), while for beam systems they tend to slowly diverge (figure 9). Again, this behaviour is linked to that of the model overlap factor, which in this case remains constant for rods but reduces with increases in driving frequency for beams.

3. When ϵ is small and the damping heavy, not only is there no great variation between ensemble members, there is also little variation from frequency to frequency,

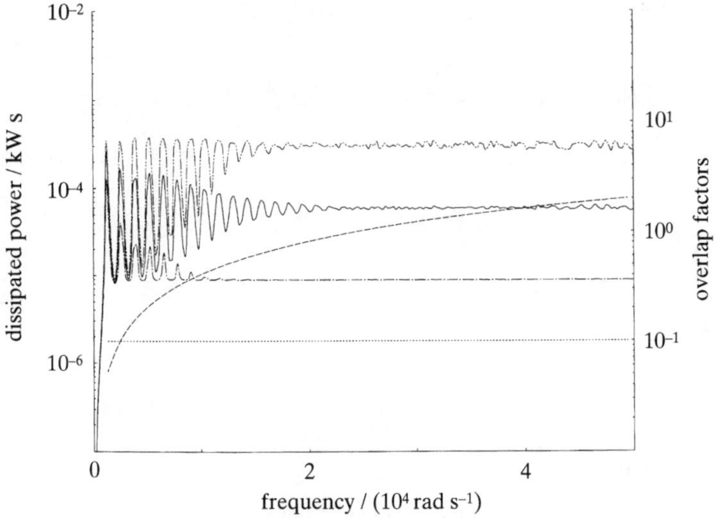

Figure 12. Statistics of the dissipated power spectral density function for rod 1 when it is *uncoupled* from rod 2; rain-on-the-roof excitation; $\epsilon = 0.05$; $c = 130$ s^{-1}; key as in figure 5.

since there are then no significant resonant peaks either. In such cases the 5% and 95% points and the mean all tend to constant values with respect to frequency, see figure 10.

4. The frequency beyond which the mean becomes stationary (non-oscillatory) reduces with increases in system randomness (cf. figures 5 and 8). The oscillatory behaviour of the power spectra is caused by two effects, namely, the occurrence of resonances and the variation of mode shape values at the points of coupling (and driving for point forcing). Clearly, in an SEA context, the frequency beyond which the effects of individual natural frequencies and mode shapes cease to dominate variations in the mean value is of considerable interest. A useful criterion for identifying this frequency can be stated in terms of the statistical overlap factor. This survey has shown that for systems under rain-on-the-roof type excitations, choosing a frequency beyond which S_n is greater than 2 guarantees steady mean behaviour (figures 5 and 8). For systems under point forcing, where the mode shapes at the driving point also enter the calculations, S_n needs to be greater than 3 for steady mean behaviour. It may be noted in this context that for the beam systems covered by this survey the statistical overlap factor never reaches such values and therefore the corresponding mean values always remained oscillatory (figures 7 and 9). If ϵ is increased to 0.20, S_n does then become high enough and the mean becomes steady for $S_n > 2$, see figure 11. It may also be noted in this context that the cutoff value of S_n is dependent on coupling strength and is smaller for weakly coupled systems. This point is illustrated in figure 12, where the power dissipated in rod 1, when it is uncoupled from the second subsystem, is shown (i.e. $k_c = 0$). In this case it is seen that $S_n > 1$ is sufficient to guarantee smooth mean behaviour. This is indicative of the fact that the variations in spectra caused by mode shape effects persists longer than those due to resonance effects.

5. For the same level of damping, larger ϵ implies smoother statistics, with the width of the confidence band reaching an upper limit (figures 9 and 11). This feature is more clearly seen in figure 13 which shows the statistics of power dissipated in the beam system under rain-on-the-roof excitation at $\omega = 25000$ rad s^{-1}, as a function of

ϵ. Clearly, when ϵ is large enough the response at any given frequency may vary all the way from a resonance to an anti-resonance and the 5% and 95% points are then directly related to the bounds of the deterministic spectra. This points towards the usefulness of the study of bounds of the power spectra for connected systems in the context of SEA response variability. In particular, a study of the statistics of the bounds on the spectra would be of interest. Note however, that the models studied here do not exhibit mode shape variations as the mass densities are always taken to be constant along the lengths of the subsystems. In situations where this is not the case, the bounds of the deterministic spectra can no longer be relied upon in this way (see Keane & Manohar 1993). It should also be noted that ϵ is greater than 0.1 before such behaviour is seen in figure 13, corresponding to a normalized standard deviation in the subsystem masses of 10%. This limiting value of ϵ was observed to be smaller for higher values of the driving frequency and subsystem modal overlap factors. On the other hand, for a given level of ϵ, greater damping was seen to result in narrower confidence bands.

6. For the constant bandwidth model, the contour of 5% probability reaches steady state faster than the contour of 95% probability (figures 5 and 8). This feature arises because the peak responses are caused by resonant behaviour and therefore, these responses can be expected to be dominated by a single mode. Conversely, the minimum responses occur at frequencies away from the resonances and therefore, where more than one mode contributes to such responses. At higher driving frequencies the overlap in the PDFs of the natural frequencies increases and the response, even at resonances, will have contributions from several neighbouring modes. This eventually leads to the stationarity of the PDF as a whole.

5. Empirical distribitions

The simulation procedure used to carry out the survey described in the preceding section is general in scope and estimates for the moments and probability distribution functions of the response power spectra can be obtained with equal ease. However, to obtain reliable estimates of the 5% and 95% probability points, a large sample size needs to be used in simulation work. In practical contexts, such large samples are seldom available and decisions often need to be made with a more limited data, where the direct estimates of 5% and 95% probability points will not be sufficiently accurate. Thus, it is desirable to develop empirical procedures to estimate these probability points using knowledge of the first few moments, which, perhaps can be estimated with relatively less difficulty. Clearly, this is a form of curve fitting and its success depends upon the choice of the PDF used to fit the data. It may be noted in this context that the nonlinear nature of the transformations of random variables implicit in equations (1)–(11) rules out the use of analytical procedures for determining the PDFs of the power spectra. Besides, the expressions for the spectra, especially for higher values of k_c, do not easily suggest any limiting distributions which might arise as $\omega \to \infty$ or as the number of terms contributing to the modal summations becomes large. Consequently, the choice of the distribution function has to be based on trial and error procedures and be guided by mathematical and computational expediency.

The experience gained by us in this context has shown that distributions with one parameter, such as the Rayleigh and exponential distributions, do not fit the data well. Amongst distributions with two parameters, the gaussian distribution was

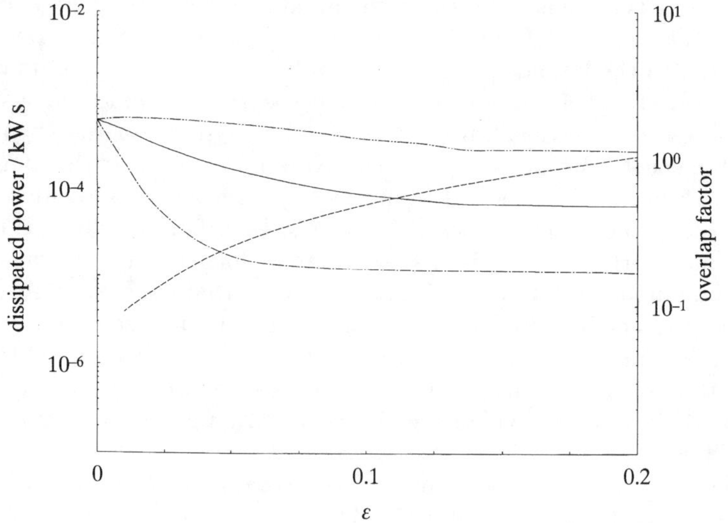

Figure 13. Statistics of the dissipated power spectral density function for subsystem 1 of the beam system; rain-on-the-roof excitation; $\omega = 25000$ rad s^{-1}; $c = 262$ s^{-1}; ——, mean; —·—·—, 50% probability point; —··—··—, 95% probability point; -------, statistical overlap factor.

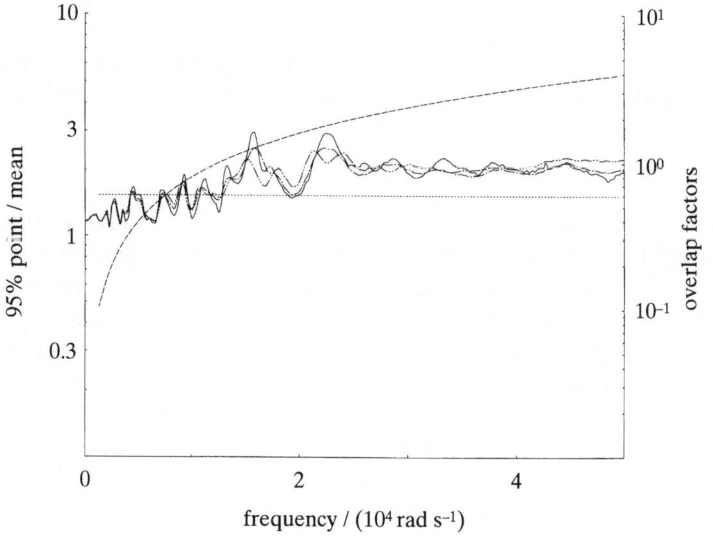

Figure 14. Empirical predictions for the 95% probability point of the dissipated power spectral density function for subsystem 1 of the rod system; point harmonic excitation; $\epsilon = 0.10$; $c = 780$ s^{-1}; ——, simulated; —·—, gamma; —··—, lognormal; ------, modal overlap factor; ——, statistical overlap factor.

found to be unsuitable, while the lognormal and gamma distributions give reasonably good fits. To assess the usefulness of these empirical distributions in describing the data, the predictions for the 5% and 95% probability points, skewness and kurtosis coefficients, as functions of ω, were compared with the corresponding simulated results. Two of the predictions for the 95% probability points are shown in figures 14 and 15. Additionally, comparisons of the probability distribution functions at a fixed driving frequency of 40000 rad s^{-1} have been given

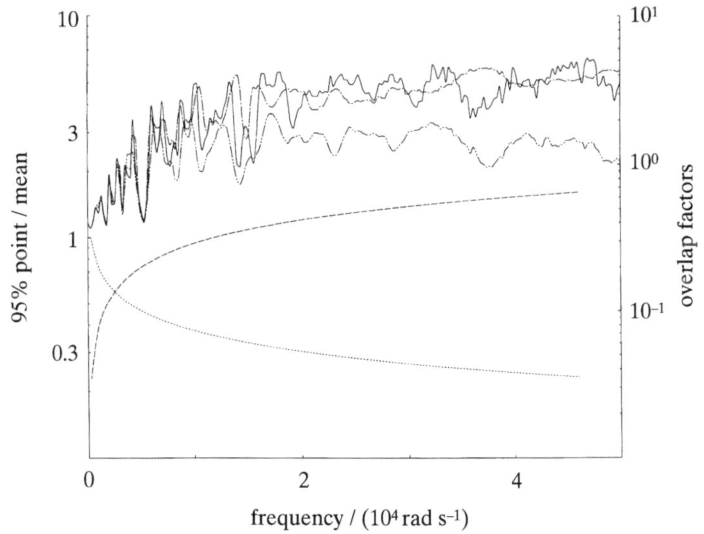

Figure 15. Empirical predictions for the 95% probability point of the dissipated power spectral density function for subsystem 1 of the beam system; rain-on-the-roof excitation; $\epsilon = 0.10$; $c = 262\ \mathrm{s}^{-1}$; key as in figure 14.

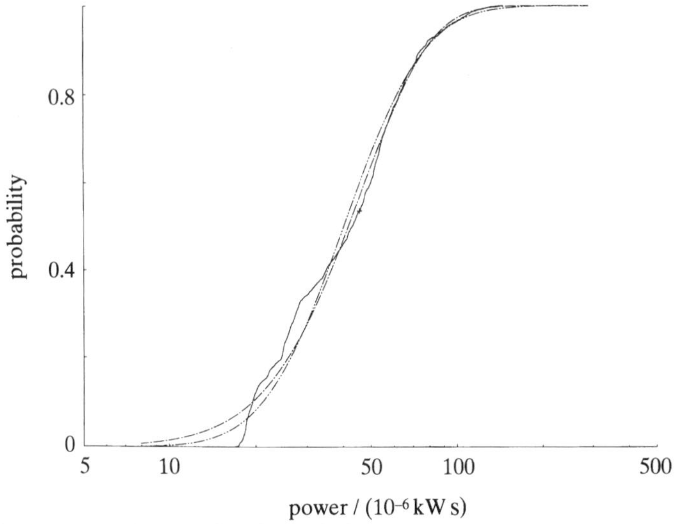

Figure 16. Empirical predictions for the probability distribution of the dissipated power spectral density function for subsystem 1 of the rod system; point harmonic excitation; $\omega = 40000$ rad s^{-1}; $\epsilon = 0.10$; $c = 780\ \mathrm{s}^{-1}$; ——, simulated; —·—, gamma; —··—, lognormal.

in figures 16 and 17. A study of these figures reveals that both the gamma and lognormal distributions provide fairly accurate estimates for the 95% probability points, although where the statistical overlap factor is low, the gamma distribution seems to give better results (with the kurtosis and skewness coefficients being generally better predicted). Conversely, the lognormal distribution works better for the 5% points.

As has already been stated, the choice of gamma and lognormal distribution in this study is arbitrary. It may be recalled that the lognormal distribution generally arises as a limiting distribution of products of independent random variables. On the other

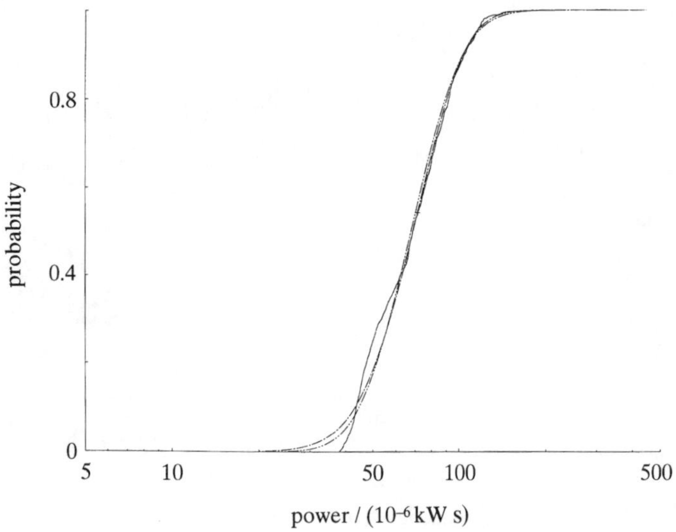

Figure 17. Empirical predictions for the probability distribution of the dissipated power for subsystem 1 of the beam system; rain-on-the-roof excitation; $\omega = 40000$ rad s^{-1}; $\epsilon = 0.10$; $\zeta = 0.03$; key as in figure 16.

hand, the gamma distribution arises in Poisson process theory as the time required for the nth success (see Benjamin & Cornell 1970). Obviously, there is no indication in the expressions for the power spectra (equations (1)–(10)) that the conditions favourable to the existence of either distribution are satisfied. Further work clearly needs to be done to place the use of these distributions on a firmer basis.

6. Conclusions

This study has considered the probability distributions of the dissipated power spectra in a system of two spring-coupled, one-dimensional subsystems. The effects of changing various parameters of the problem on the behaviour of the PDFs of the response spectra have been studied using Monte Carlo simulation procedures. These changes have encompassed choice of subsystem type, damping model, strength of system randomness and type of excitation (an earlier study considered various classes of subsystem randomness (see Keane & Manohar 1993)). The present work pays particular attention to the choice of damping model and the strength of system randomness. The effects of these quantities have been characterized in terms of the modal overlap factor and a newly introduced statistical overlap factor, S_n. The modal overlap factor takes into account the effect of subsystem type and damping while the statistical overlap factor reflects details of the system randomness.

The work presented shows that a cutoff frequency beyond which the mean response spectra become stationary can be determined by reference to the statistical overlap factor. For the cases considered here, systems driven by point harmonic forcing have mean responses independent of frequency when S_n is greater than 3, while for rain-on-the-roof excitations S_n need only be greater than 2. For systems with weak coupling, it turns out that the above conditions can be relaxed to $S_n > 2$ and $S_n > 1$. This study has also demonstrated that driving frequency dependent variations in the 5% and 95% probability points are strongly influenced by corresponding variations in the modal overlap factor. For rod systems with constant bandwidth damping models, where the modal overlap factor remains constant with driving frequency, the probability points tend to a constant spacing. For beam systems with constant bandwidth damping models, where the model overlap factor reduces with frequency, the probability points are observed to slowly diverge from each other. Conversely, for constant ζ_n damping models, where the modal overlap factors for both rod and beam systems rise with increases in driving frequency, the probability points converge towards the mean.

In summary, it may be said that, if the modal and statistical overlap factors are both large (> 3, say), the responses of ensemble members tend to show moderate deviations from the mean while the mean remains sensibly independent of driving frequency (e.g. significant damping and parameter randomness, see figure 6). If only the modal overlap factor is large, there will be small deviations from the mean and little variation from frequency to frequency (e.g. significant damping with little parameter randomness, see figure 10). Conversely, if only the statistical overlap factor is large, the 5% and 95% probability points tend to be widely spaced but again the mean tends not to vary much from frequency to frequency (e.g. light damping with significant parameter randomness, see figure 8). Finally, if neither overlap factor is large, although the 5% and 95% points may not be widely spaced the mean is likely to show violent variations from frequency to frequency (e.g. light damping with small parameter randomness, see figure 5). It is usually this last case that is of most interest in structural dynamics and so, since both overlap factors can vary with frequency, precise knowledge of their behaviour would seem to be a precursor to the successful application of SEA methods. Unfortunately, although typical modal overlap factors may be found from a single realization of a problem, the statistical factor requires information from a potentially large population. This may well be available from the output of industrial production lines; it is less easy to derive when dealing with complicated built up structures made in small batches. In the absence of such data the maximum and minimum values observed in the spectra of a single realization may be used to *approximate* the 5% and 95% confidence limits for the ensemble across the range of frequencies surrounding such points; however, such an approach will tend to place the 5% and 95% points further from the mean than is likely in most real structures. Moreover, they will be unable to reflect situations where unusual mode shapes may arise, such as in nearly periodic structures. Alternatively, the 5% and 95% probability points may be estimated using knowledge of the first two moments of the response based on a small sample of systems and an assumed probability function. Preliminary investigations have shown that the lognormal and gamma probability distributions can usefully be used for this purpose, although questions still remain on justifying the choice of these distributions.

This work was supported by funding from the U.K. Department of Trade and Industry which is gratefully acknowledged.

References

Benjamin, J. R. & Cornell, C. A. 1970 *Probability, statistics and decision for civil engineers*. New York: McGraw-Hill.

Craik, R. J. M., Steel, J. A. & Evans, D. I. 1991 Statistical energy analysis of structure-borne sound transmission at low frequencies. *J. Sound Vib.* **144**, 95–107.

Davies, H. G. 1973 Random vibration of distributed systems strongly coupled at discrete points. *J. Acoust. Soc. Am.* **54**, 507–515.

Davies, H. G. & Wahab, M. A. 1981 Ensemble averages of power flow in randomly excited coupled beams. *J. Sound Vib.* **77**, 311–321.

Davies, H. G. & Khandoker, S. I. 1982 Random point excitation of coupled beams. *J. Sound Vib.* **84**, 557–562.

Fahy, F. J. & Mohammed, A. D. 1992 A study of uncertainty in applications of SEA to coupled beam and plate systems, part I: Computational experiments. *J. Sound Vib.* **158**, 45–67.

Heron, K. 1990 The development of a wave approach to statistical energy analysis. *Proc. IOA* **12**, 551–555.

Hodges, C. H. & Woodhouse, J. 1986 Theories of noise and vibration transmission in complex structures. *Rep. Prog. Phys.* **49**, 107–170.

Keane, A. J. 1992 Energy flows between arbitrary configurations of conservatively coupled multi-modal elastic subsystems. *Proc. R. Soc. Lond.* A **436**, 537–568.

Keane, A. J. & Manohar, C. S. 1993 Energy flow variability in a pair of coupled stochastic rods. *J. Sound Vib.* **168**, 253–284.

Lin, Y. K. 1967 *Probabilistic theory of structural dynamics.* McGraw-Hill.

Lyon, R. H. 1969 Statistical analysis of power injection and response in structures and rooms. *J. Acoust. Soc. Am.* **45**, 545–565.

Lyon, R. H. 1975 *Statistical energy analysis of dynamical systems: theory and applications.* MIT Press.

Lyon, R. H. & Eichler, E. 1964 Random vibration of connected structures. *J. Acoust. Soc. Am.* **36**, 1344–1354.

Scharton, T. D. & Lyon, R. H. 1968 Power flow and energy sharing in random vibration. *J. Acoust. Soc. Am.* **43**, 1332–1343.

Skudrzyk, E. 1968 *Simple and complex vibratory systems.* Pennsylvania State University Press.

Skudrzyk, E. 1980 The mean value method of predicting the dynamic response of complex vibrators. *J. Acoust. Soc. Am.* **67**, 1105–1135.

Skudrzyk, E. 1987 Understanding the dynamic behaviour of complex vibrators. *Acoustica* **64**, 123–147.

Soong, T. T. & Bogdanoff, J. L. 1963 On the natural frequencies of a disordered linear chain of N degrees of freedom. *Int. J. Mech. Sci.* **5**, 237–265.

Statistical energy analysis of nonconservatively coupled systems

By M. Beshara[1], G. Y. Chohan[2], A. J. Keane[1] and W. G. Price[2]

[1]*Department of Engineering Science, University of Oxford, Parks Road, Oxford, OX1 3PJ, England*

[2]*Department of Ship Science, University of Southampton, Southampton, SO17 1BJ, England*

The axial vibrations of two rods coupled together by a spring and viscous damper are investigated using a modal analysis approach. Exact expressions describing various energy flows are derived in terms of external forcing spectra. The effects of changes to coupling parameters on the relevant power receptances are studied and attention forcused on analysis of the coupling damper. The conditions when significant power is dissipated within this damper are examined, thus highlighting when the nonconservative nature of the coupling cannot be neglected without major error. Finally relationships between the ensemble avarage energy flow and the average total energies of the subsystems are recovered. The basic features of these various relationships are illustrated throughout by the use of numerical examples.

1. Introduction

Statistical energy analysis (SEA) is a tool that may be used for the analysis of complex systems when the usual deterministic methods, such as finite element analysis, are no longer practical. It is used mainly when dealing with complex systems at high frequencies, where the large number of degrees of freedom required for finite element methods lead to dramatically increased computing times and costs, even with powerful computational facilities. Moreover, at high frequencies, the results obtained by deterministic methods suffer from some shortcomings. These arise because of uncertainties in material properties plus the sensitivity of mode shapes and modal resonant frequencies to any changes in boundary conditions or damping distribution, which lead to significant differences in the results obtained for nearly identical structures at such frequencies. It is these uncertainties that require the treatment of high frequency dynamic response prediction as a probabilistic problem. Usually, an ensemble of similar systems is considered which differ in their parameters, and an ensemble average of the response is then predicted, often that of the total energies of the subsystems, which are then used to represent the responses of these subsystems. Engineering applications of SEA normally involve the analysis of complex systems (e.g., buildings designed against earthquakes, jet engines, aerospace structures, ships, etc.). It is standard SEA practice to divide these systems into sets of subsystems, often described by their gross dynamic properties, which receive, dissipate and exchange energy, see for example, Norton (1989). The concept of using energy flows to describe the interaction between such subsystems was first proposed by Lyon and Maidanik (1962), see also Scharton and Lyon (1968) and Lyon (1975).

In the traditional SEA approach, the system is modelled into subsystems with weak conservative coupling, the aim being to estimate the total time-average distribution of energy among the coupled subsystems. The evaluation of the coefficients that relate the energy flows between subsystems to their energy levels thus lies at the heart of SEA. It is extensively discussed by Lyon (1975), where the energy flow between two oscillators coupled by a spring is analysed, leading to the basic principle of SEA, which states that average energy flow is proportional to the difference between the total time averaged energies of the coupled oscillators.

Since that time, extensive studies have focused on the analysis of two coupled multi-modal systems, as many different types of structures may be idealized in this way. Energy flows in beams have been studied by Crandall and Lotz (1971), Goyder *et al.* (1956), Mace (1993) and Fahy and Mohammed (1992) using wave propagation approaches. Remington and Manning (1975) studied the energy flow between two coupled rods using the same approach and compared the results with exact solutions derived using Green functions. Davies (1972*b*) used the modal approach for the analysis of two coupled beams, Keane and Price (1991) and Keane (1988) used the same approach for the analysis of coupled rods. Their results agree with those derived by Remington and Manning (1975). In all these studies, the energy flow equations were derived assuming a conservative coupling and it was found that, in general, the energy flow between two coupled multi-modal subsystems is proportional to the difference in their modal energies (given a number of assumptions which place limitations on the validity of SEA when dealing with certain kinds of problems; these various assumptions and limitations are extensively discussed in many of the cited works, see for example, Keane (1988) or Fahy (1974)).

When the coupling mechanism between two subsystems is nonconservative, it is clear that the standard proportionality relationship is no longer valid. This problem appears to have been investigated first by Lyon (1975) for coupled oscillators, then by Fahy and Yao (1987), Sun, Lalor and Richards (1987) and Chen and Soong (1991) where it was shown that the energy flow between the two oscillators depends also on the sum of the energy levels of the two oscillators, not just their difference.

In the work presented here, the axial vibrations of two rods coupled together by both a spring and a viscous damper are investigated, using modal analysis. Following steps similar to Davies (1972*a*) and Keane (1988), deterministic expressions for the input powers, dissipated powers and energy flow between the subsystems are derived. By introducing a complex coupling stiffness, Ω, which includes both the spring and damper strengths, it is found that the results obtained agree with those derived by Keane (1988) for conservative coupling, except that a complex coupling strength must be used in place of a simple spring constant. Next, two forcing models are considered which allow the separation of frequency and spatial variables in the expressions of modal spectra. The effects of changes in the coupling parameters on the various receptances are then considered, and special attention paid to the power dissipated within the coupling. This is found to remain at relatively low levels except for a specific range of values of the coupling parameters. Moreover, it is shown that the power dissipated within the coupling is relatively small compared with the power transferred through it when dealing with couplings in this range, which suggests that the nonconservative nature of the coupling can often be ignored without introducing significant errors. Finally, a relationship linking the energy flow between the subsystems and their total energy levels is recovered. As expected, the energy flow between the two rods is dependent not only on the difference between the average modal energies of the two subsystems, but also on the sum of their energy

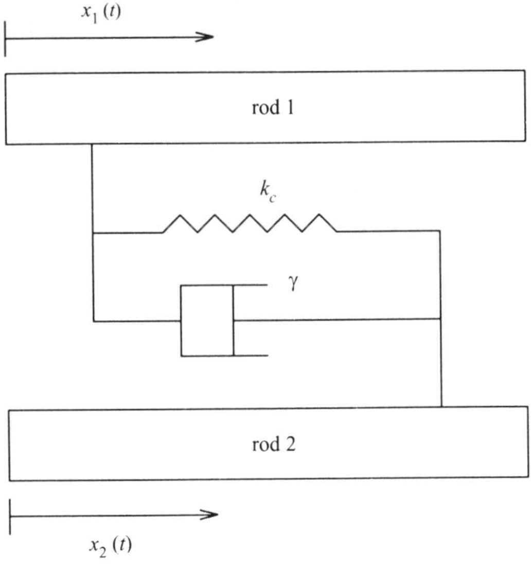

Figure 1. Two point-coupled axially vibrating rods.

levels. This result agrees with that derived by Chen and Soong (1991) for two coupled oscillators, although the constants of proportionality are, of course, different.

2. Derivation of modal summation results

Consider a multi-modal system comprising of two free-free rods coupled together at $x_1 = a_1$ and $x_2 = a_2$ by a spring of stiffness k_c and a damper of strength γ, as is illustrated in figure 1. The mass densities per unit length are $\rho_1(x_1)$ and $\rho_2(x_2)$ and the rigidities per unit length $EA_1(x_1)$ and $EA_2(x_2)$, for rods 1 and 2, respectively. Damping within the two subsystems is assumed proportional to the density per unit length, so that the damping constants for the subsystems, $r_1(x_1)$ and $r_2(x_2)$, are given by,

$$r_1(x_1) = c_1\rho_1(x_1) \quad \text{and} \quad r_2(x_2) = c_2\rho_2(x_2). \tag{2.1,2.2}$$

The differential equations of motion which govern the dynamic behaviour of the rods are then

$$\{\rho_1(x_1)\partial^2/\partial t^2 + r_1(x_1)\partial/\partial t + \Lambda_1\}y_1(x_1,t) = P_1(x_1,t) \tag{2.3}$$

and

$$\{\rho_2(x_2)\partial^2/\partial t^2 + r_2(x_1)\partial/\partial t + \Lambda_2\}y_2(x_2,t) = P_2(x_2,t) \tag{2.4}$$

where Λ_1, Λ_2 are linear spatial differential operators on subsystems 1 and 2, and $P_1(x_1,t)$ and $P_2(x_2,t)$ are the forcing functions on the subsystems. These are given by the expressions

$$P_1(x_1,t) = F_1(x_1,t) + \{k_c[y_2(a_2,t) - y_1(a_1,t)] + \gamma[\dot{y}_2(a_2,t) - \dot{y}_1(a_1,t)]\}\delta(x_1 - a_1) \tag{2.5}$$

and

$$P_2(x_2,t) = F_2(x_2,t) + \{k_c[y_1(a_1,t) - y_2(a_2,t)] + \gamma[\dot{y}_1(a_1,t) - \dot{y}_2(a_2,t)]\}\delta(x_2 - a_2). \tag{2.6}$$

This problem may be solved using modal analysis where the characteristic equations of the subsystems, when uncoupled, are given in terms of their mode shapes Ψ_i and natural frequencies ω_i with

$$\Lambda\Psi_i(x) - \rho(x)\,\omega_i^2\Psi_i(x) = 0. \tag{2.7}$$

This classical eigenvalue problem is readily solved to gain the natural modes and natural frequencies of the two subsystems, see for example Meirovitch (1975). The natural modes are then normalised to satisfy the orthogonality conditions,

$$\int_1 \Psi_i(x_1)\Psi_j(x_1)\rho_1(x_1)\mathrm{d}x_1 = M_1\delta_{ij} \quad \text{and} \quad \int_2 \Psi_r(x_2)\Psi_s(x_2)\rho_2(x_2)\mathrm{d}x_2 = M_2\delta_{rs} \tag{2.8,2.9}$$

where M_1 and M_2 are the total masses of subsystems 1 and 2, respectively. In the following analysis i, j refer to the first subsystem while r, s refer to the second subsystem. According to the expansion theorem, the displacements of the two coupled subsystems can be written as an expansion in terms of the natural modes with the modal components $W_i(t)$ and $W_r(t)$ so that:

$$y_1(x_1,t) = \sum_{i=1}^{\infty} \Psi_i(x_1)W_i(t) \quad \text{and} \quad y_2(x_2,t) = \sum_{i=1}^{\infty} \Psi_r(x_2)W_i(t). \tag{2.10,2.11}$$

The modal components of the external forcing are given in terms of the natural modes by the expressions:

$$L_i(t) = \int_1 F_1(x_1,t)\Psi_i(x_1)\mathrm{d}x_1 \quad \text{and} \quad L_r(t) = \int_2 F_2(x_2,t)\Psi_r(x_2)\mathrm{d}x_2, \tag{2.12,2.13}$$

and the modal components of the coupling forces are given by the expressions

$$V_i(t) = \int_1 \{k_c[y_2(a_2,t) - y_1(a_1,t)] + \gamma[\dot{y}_2(a_2,t) - \dot{y}_1(a_1,t)]\}\delta(x_1 - a_1)\Psi_i(x_1)\mathrm{d}x_1 \tag{2.14}$$

and

$$V_r(t) = \int_2 \{k_c[y_1(a_1,t) - y_2(a_2,t)] + \gamma[\dot{y}_1(a_1,t) - \dot{y}_2(a_2,t)]\}\delta(x_2 - a_2)\Psi_r(x_2)\mathrm{d}x_2 \tag{2.15}$$

so that the external forcing may be written in terms of their modal components as

$$P_1(x_1,t) = \sum_i \Psi_i(x_1)\frac{\rho_1(x_1)}{M_1}[L_i(t) + V_i(t)] \tag{2.16}$$

and

$$P_2(x_2,t) = \sum_r \Psi_r(x_2)\frac{\rho_2(x_2)}{M_2}[L_r(t) + V_r(t)]. \tag{2.17}$$

Considering equations(2.10–2.17) and substituting into the two equations of motion leads to

$$\rho_1(x_1)\sum_i \Psi_i(x_1)[\ddot{W}_i(t) + c_1\dot{W}_i(t) + \omega_i^2 W_i(t)] = \sum_i \Psi_i(x_1)\frac{\rho_1(x_1)}{M_1}[L_i(t) + V_i(t)] \tag{2.18}$$

and

$$\rho_2(x_2)\sum_r \Psi_r(x_2)[\ddot{W}_r(t) + c_2\dot{W}_r(t) + \omega_r^2 W_r(t)] = \sum_r \Psi_r(x_2)\frac{\rho_2(x_2)}{M_2}[L_r(t) + V_r(t)] \tag{2.19}$$

Multiplying both sides of equation (2.18) by $\Psi_j(x_1)$ and both sides of equation (2.19) by $\Psi_s(x_2)$ and taking the integral over the range $-\infty$ to ∞ yields:

$$M_1[\ddot{W}_j(t) + c_1\dot{W}_j(t) + \omega_j^2 W_j(t)] = [L_j(t) + V_j(t)] \tag{2.20}$$

and

$$M_2[\ddot{W}_s(t) + c_2\dot{W}_s(t) + \omega_s^2 W_s(t)] = [L_s(t) + V_s(t)]. \tag{2.21}$$

These last two equations are in the time domain: to proceed further in the analysis both of them are written in the frequency domain by taking Fourier transforms of each equation. Hence the first equation becomes

$$\begin{aligned} M_1(-\omega^2 + \omega_j^2 + \mathrm{i}c_1\omega)W_j(\omega) &= L_j(\omega) + k_c\Psi_j(a_1)[\textstyle\sum_r \Psi_r(a_2)W_r(\omega) - \sum_i \Psi_i(a_1)W_i(\omega)] \\ &\quad + \mathrm{i}\gamma\omega\Psi_j(a_1)[\textstyle\sum_r \Psi_r(a_2)W_r(\omega) - \sum_i \Psi_i(a_1)W_i(\omega)]. \end{aligned} \tag{2.22}$$

Defining $\Phi_j(\omega)$ and $\Phi_r(\omega)$ as

$$\Phi_j = (-\omega^2 + \omega_j^2 + \mathrm{i}c_1\omega) \quad \text{and} \quad \Phi_r = (-\omega^2 + \omega_r^2 + \mathrm{i}c_2\omega) \tag{2.23,2.24}$$

and the complex coupling strength as

$$\Omega = k_c + \mathrm{i}\gamma\omega, \tag{2.25}$$

the equations become

$$M_1\Phi_j(\omega)W_j(\omega) = L_j(\omega) + \Omega\Psi_j(a_1)[\sum_r \Psi_r(a_2)W_r(\omega) - \sum_i \Psi_i(a_1)W_i(\omega)] \tag{2.26}$$

and

$$M_2\Phi_s(\omega)W_s(\omega) = L_s(\omega) + \Omega\Psi_s(a_2)[\sum_i \Psi_i(a_1)W_i(\omega) - \sum_r \Psi_r(a_2)W_r(\omega)]. \tag{2.27}$$

Multiplying both sides of equation (2.26) by $\frac{\Psi_j(a_1)}{M_1\Phi_j(\omega)}$ and both sides of equation (2.27) by $\frac{\Psi_j(a_1)}{M_1\Phi_j(\omega)}$ and taking the summation over j and s, respectively, the two equations then have the form

$$\begin{aligned} \sum_j \Psi_j(a_1)W_j(\omega) &= \sum_j \frac{L_j(\omega)\Psi_j(a_1)}{M_1\Phi_j(\omega)} \\ &\quad + \Omega\sum_j \frac{\Psi_j^2(a_1)}{M_1\Phi_j(\omega)}\left[\sum_r \Psi_r(a_2)W_r(\omega) - \sum_i \Psi_i(a_1)W_i(\omega)\right] \end{aligned} \tag{2.28}$$

and

$$\begin{aligned} \sum_s \Psi_s(a_2)W_s(\omega) &= \sum_s \frac{L_s(\omega)\Psi_s(a_2)}{M_2\Phi_s(\omega)} \\ &\quad + \Omega\sum_s \frac{\Psi_s^2(a_2)}{M_2\Phi_s(\omega)}\left[\sum_i \Psi_i(a_1)W_i(\omega) - \sum_r \Psi_r(a_2)W_r(\omega)\right]. \end{aligned} \tag{2.29}$$

These two simultaneous equations can be solved for the various summations to give

$$\begin{aligned} \sum_j \Psi_j(a_1)W_j(\omega) &= \sum_j \frac{L_j(\omega)\Psi_j(a_1)}{M_1\Phi_j(\omega)} \\ &\quad + \frac{\Omega}{\Delta}\sum_j \frac{\Psi_j^2(a_1)}{M_1\Phi_j(\omega)}\left[\sum_r \frac{L_r\Psi_r(a_2)}{M_2\Phi_r(\omega)} - \sum_i \frac{L_i(\omega)\Psi_i(a_1)}{M_1\Phi_i(\omega)}\right] \end{aligned} \tag{2.30}$$

and

$$\sum_s \Psi_s(a_2)W_s(\omega) = \sum_s \frac{L_s(\omega)\Psi_s(a_2)}{M_2\Phi_s(\omega)} + \frac{\Omega}{\Delta}\sum_s \frac{\Psi_s^2(a_2)}{M_2\Phi_s(\omega)}\left[\sum_i \frac{L_i\Psi_i(a_1)}{M_1\Phi_i(\omega)} - \sum_r \frac{L_r(\omega)\Psi_r(a_2)}{M_2\Phi_r(\omega)}\right] \tag{2.31}$$

where

$$\Delta = 1 + \Omega\left[\sum_j \frac{\Psi_j^2(a_1)}{M_1\Phi_j(\omega)} + \sum_s \frac{\Psi_s^2(a_2)}{M_2\Phi_s(\omega)}\right]. \tag{2.32}$$

As these equations hold irrespective of the individual natural frequencies and modes shapes, it is then possible to take just the jth term of the summation in equation (2.30) to give

$$W_j(\omega) = \frac{[\kappa_j(\omega) + L_j(\omega)]}{M_1\Phi_j(\omega)} \tag{2.33}$$

where

$$\kappa_j = \frac{\Omega\Psi_j(a_1)}{\Delta}\left[\sum_r \frac{L_r(\omega)\Psi_r(a_2)}{M_2\Phi_r(\omega)} - \sum_i \frac{L_i(\omega)\Psi_i(a_1)}{M_1\Phi_i(\omega)}\right]. \tag{2.34}$$

This result is equivalent to equation 7 of Davies' work (1972*a*) and equation A29 of Keane's work (1988).

3. Long term averages

In the following sections attention is focused on the calculation of energy flows which involve the products of various time-varying functions (forces, displacements and velocities). Before proceeding to the derivation of the energy flows, it is useful to summerize briefly the relations between the energy flow through the coupling and the power dissipated within it.

The time-average energy flow from rod 1 into the coupling is here denoted by $\langle\Pi'_{12}\rangle$. Since the coupling mechanism is not conservative, part of this power will be dissipated within the coupling and this time-average dissipated power is denoted by $\langle\Pi_{dc}\rangle$. The power transferred to rod 2 is then simply calculated by subtracting the power dissipated in the coupling from the total power that leaves rod 1, i.e., the time-average power transferred to rod 2 is given by

$$\langle\Pi_{12}\rangle = \langle\Pi'_{12}\rangle - \langle\Pi_{dc}\rangle. \tag{3.1}$$

Next the time-average power flow from rod 2 into the coupling is denoted by $\langle\Pi'_{21}\rangle$ and the time-average power flow transferred from rod 2 to rod 1 by $\langle\Pi_{21}\rangle$. The relation between these two time-averages is similarly

$$\langle\Pi_{21}\rangle = \langle\Pi'_{21}\rangle - \langle\Pi_{dc}\rangle \tag{3.2}$$

and so it is obvious that

$$\langle\Pi_{12}\rangle = -\langle\Pi'_{21}\rangle \quad \text{and} \quad \langle\Pi_{21}\rangle = -\langle\Pi'_{12}\rangle. \tag{3.3,3.4}$$

The energy balance equation for the coupling is then

$$\langle\Pi_{12}\rangle + \langle\Pi_{21}\rangle + \langle\Pi_{dc}\rangle = 0 \tag{3.5}$$

or

$$\langle \Pi'_{12} \rangle + \langle \Pi'_{21} \rangle = \langle \Pi_{dc} \rangle. \tag{3.6}$$

Also the energy balance equations for subsystems 1 and 2 are

$$\langle \Pi_{1_{in}} \rangle - \langle \Pi_{1_{diss}} \rangle - \langle \Pi'_{12} \rangle = 0 \quad \text{and} \quad \langle \Pi_{2_{in}} \rangle - \langle \Pi'_{2_{diss}} \rangle - \langle \Pi'_{21} \rangle = 0 \qquad (3.7,3.8)$$

Since all the processes considered here are assumed ergodic, the time averages taken for any one system will be equal to the ensemble averages taken across an infinite set of similar systems, i.e., all the time averages $\langle\ \rangle$ can be replaced by ensemble averages $\mathrm{E}[\,]$.

(a) Energy flow

Consider now the energy flow Π_{21} from rod 1 to rod 2. It is given in the time domain by the expression

$$\begin{aligned} \Pi_{21}(t) &= \int_1 \{k_c[y_2(a_2,t) - y_1(a_1,t)] + \gamma[\dot{y}_2(a_2,t) - \dot{y}_1(a_1,t)]\}\delta(x_1 - a_1)\dot{y}_1(x_1,t)\mathrm{d}x_1 \\ &= \dot{y}_1(a_1,t)\{k_c[y_2(a_2,t) - y_1(a_1,t)] + \gamma[\dot{y}_2(a_2,t) - \dot{y}_1(a_1,t)]\}, \end{aligned} \tag{3.9}$$

and the ensemble average of this energy is

$$\begin{aligned} \mathrm{E}[\Pi_{21}(t)] = \mathrm{E}[\Pi_{21}] = k_c\{&\mathrm{E}[\dot{y}_1(a_1,t)y_2(a_2,t)] - \mathrm{E}[\dot{y}_1(a_1,t)y_1(a_1,t)]\} \\ &+\gamma\{\mathrm{E}[\dot{y}_1(a_1,t)\dot{y}_2(a_2,t)] - \mathrm{E}[\dot{y}_1^2(a_1,t)]\}. \end{aligned} \tag{3.10}$$

Since $\mathrm{E}[\dot{y}_1(a_1,t)y_1(a_1,t)] = 0$, see Newland (1975), this last expression becomes

$$\mathrm{E}[\Pi_{21}] = k_c\{\mathrm{E}[\dot{y}_1(a_1,t)y_2(a_2,t)]\} + \gamma\{\mathrm{E}[\dot{y}_1(a_1,t)\dot{y}_2(a_2,t)] - \mathrm{E}[\dot{y}_1^2(a_1,t)]\}. \tag{3.11}$$

Inserting the modal expansions for the displacements leads to

$$\begin{aligned} \mathrm{E}[\Pi_{21}] = k_c \sum_i \sum_r \Psi_i(a_1)\Psi_r(a_2)\mathrm{E}[\dot{W}_i(t)W_r(t)] + \gamma \sum_i \sum_r \Psi_i(a_1)\Psi_r(a_2)\mathrm{E}[\dot{W}_i(t)\dot{W}_r(t)] \\ - \gamma \sum_i \sum_j \Psi_i(a_1)\Psi_j(a_1)\mathrm{E}[\dot{W}_i(t)\dot{W}_j(t)]. \end{aligned} \tag{3.12}$$

Writing this last equation in the frequency domain then gives

$$\begin{aligned} \Pi_{21}(\omega) = k_c \sum_i \sum_r \Psi_i(a_1)\Psi_r(a_2)S_{W_iW_r}(\omega) + \gamma \sum_i \sum_r \Psi_i(a_1)\Psi_r(a_2)S_{W_iW_r}(\omega) \\ - \gamma \sum_i \sum_j \Psi_i(a_1)\Psi_j(a_1)S_{\dot{W}_i\dot{W}_r}(\omega) \end{aligned} \tag{3.13}$$

or

$$\begin{aligned} \Pi_{21}(\omega) = \mathrm{i}k_c \sum_i \sum_r \Psi_i(a_1)\Psi_r(a_2)S_{W_iW_r}(\omega) + \gamma\omega^2 \sum_i \sum_r \Psi_i(a_1)\Psi_r(a_2)S_{W_iW_r}(\omega) \\ - \gamma\omega^2 \sum_i \sum_j \Psi_i(a_1)\Psi_j(a_1)S_{W_iW_r}(\omega). \end{aligned} \tag{3.14}$$

Next, the spectral density of the derived process $S_{W_iW_r}(\omega)$ can be replaced by its equivalent in terms of the modal components of the driving forces acting on the subsystems,

$$S_{W_iW_r}(\omega) = \lim_{T\to\infty}[W_i^*(\omega)W_r(\omega)]\frac{2\pi}{T} = \lim_{T\to\infty}\left[\frac{[L_i(\omega)+\kappa_i(\omega)]^*[L_i(\omega)+\kappa_r(\omega)]}{[M_1\Phi_i(\omega)]^*[M_2\Phi_r(\omega)]}\right]\frac{2\pi}{T} \tag{3.15}$$

Multiplying these various terms and replacing $\lim_{T\to\infty}[L_i^*(\omega)L_r(\omega)]\frac{2\pi}{T}$ by $S_{ir}(\omega)$ and making the necessary mathematical manipulations gives the desired power transmitted to subsystem 1 as

$$\begin{aligned}
\Pi_{21}(\omega) &= \frac{[-\mathrm{i}k_c\omega+\gamma\omega^2]}{M_1M_2}\sum_i\sum_r\frac{\Psi_i(a_1)\Psi_r(a_2)S_{ir}(\omega)}{\Phi_i^*(\omega)\Phi_r(\omega)} \\
&+\frac{[-\mathrm{i}k_c\omega+\gamma\omega^2]}{\Delta^*}\sum_i\frac{\Omega^*\Psi_i^2(a_1)}{M_1\Phi_i^*(\omega)}\sum_r\frac{\Psi_r^2(a_2)}{M_2\Phi_r^*(\omega)}\left[\sum_r\frac{\Psi_s^2(a_2)S_{sr}(\omega)}{M_2\Phi_s^*(\omega)}-\sum_j\frac{\Psi_j^2(a_1)S_{jr}(\omega)}{M_1\Phi_j^*(\omega)}\right] \\
&+\frac{[-\mathrm{i}k_c\omega+\gamma\omega^2]}{\Delta^*}\sum_r\frac{\Omega\Psi_r^2(a_2)}{M_2\Phi_r^*(\omega)}\sum_i\frac{\Psi_i(a_1)}{M_1\Phi_i^*(\omega)}\left[\sum_j\frac{\Psi_j(a_1)S_{ij}(\omega)}{M_1\Phi_j(\omega)}-\sum_r\frac{\Psi_r(a_2)S_{ir}(\omega)}{M_2\Phi_r(\omega)}\right] \\
&+\frac{[-\mathrm{i}k_c\omega+\gamma\omega^2]}{|\Delta|^2}\sum_i\frac{\Omega^*\Psi_i^2(a_1)}{M_1\Phi_i^*(\omega)}\sum_r\frac{\Omega\Psi_r^2(a_2)}{M_2\Phi_r(\omega)}\left[-\sum_i\sum_j\frac{\Psi_i(a_1)\Psi_j(a_1)S_{ij}(\omega)}{M_1^2\Phi_i^*(\omega)\Phi_j(\omega)}\right. \\
&\left.-\sum_r\sum_s\frac{\Psi_r(a_2)\Psi_s(a_2)S_{rs}(\omega)}{M_2^2\Phi_r^*(\omega)\Phi_s(\omega)}+\sum_i\sum_r\left[\frac{\Psi_i(a_1)\Psi_r(a_2)S_{ir}(\omega)}{M_1M_2\Phi_i^*(\omega)\Phi_r(\omega)}+\frac{\Psi_i(a_1)\Psi_r(a_2)S_{ri}(\omega)}{M_1M_2\Phi_r^*(\omega)\Phi_i(\omega)}\right]\right] \\
&-\frac{\gamma\omega^2}{M_1^2}\sum_i\sum_j\frac{\Psi_i(a_1)\Psi_j(a_1)S_{ij}(\omega)}{\Phi_i^*(\omega)\Phi_j(\omega)} \\
&-\frac{\gamma\omega^2}{\Delta^*}\sum_i\frac{\Omega^*\Psi_i^2(a_1)}{M_1\Phi_i^*(\omega)}\sum_j\frac{\Psi_j(a_1)}{M_1\Phi_j(\omega)}\left[\sum_s\frac{\Psi_s(a_2)S_{sj}(\omega)}{M_2\Phi_s^*(\omega)}-\sum_{j_1}\frac{\Psi_{j_1}(a_1)S_{j_1j}(\omega)}{M_1\Phi_{j_1}^*(\omega)}\right] \\
&-\frac{\gamma\omega^2}{\Delta}\sum_j\frac{\Omega\Psi_j^2(a_1)}{M_1\Phi_j^*(\omega)}\sum_i\frac{\Psi_i(a_1)}{M_1\Phi_i^*(\omega)}\left[\sum_r\frac{\Psi_r(a_2)S_{ir}(\omega)}{M_2\Phi_r(\omega)}-\sum_{j_1}\frac{\Psi_{j_1}(a_1)S_{ij_1}(\omega)}{M_1\Phi_{j_1}(\omega)}\right] \\
&-\frac{\gamma\omega^2}{|\Delta|^2}\sum_i\frac{\Omega^*\Psi_i^2(a_1)}{M_1\Phi_i^*(\omega)}\sum_j\frac{\Omega\Psi_j^2(a_1)}{M_1\Phi_j(\omega)}\left[\sum_i\sum_j\frac{\Psi_i(a_1)\Psi_j(a_1)S_{ij}(\omega)}{M_1^2\Phi_i^*(\omega)\Phi_j(\omega)}\right. \\
&\left.+\sum_r\sum_s\frac{\Psi_r(a_2)\Psi_s(a_2)S_{rs}(\omega)}{M_2^2\Phi_r^*(\omega)\Phi_s(\omega)}-\sum_i\sum_r\left[\frac{\Psi_i(a_1)\Psi_r(a_2)S_{ir}(\omega)}{M_1M_2\Phi_i^*(\omega)\Phi_r(\omega)}+\frac{\Psi_i(a_1)\Psi_r(a_2)S_{ri}(\omega)}{M_1M_2\Phi_r^*(\omega)\Phi_i(\omega)}\right]\right]
\end{aligned} \tag{3.16}$$

Now this complex expression can be greatly simplified if the driving forces are assumed statistically independent, so that the cross-spectral densities of their modal components

$S_{ir}(\omega) = 0$, and also by recalling the definition of Δ to get

$$\Pi_{21}(\omega) = -\mathrm{i}k_c\omega \left[\frac{\left(\sum_j \frac{\Omega \Psi_j^2(a_1)}{M_1 \Phi_j(\omega)} \right)^* + \left| \sum_j \frac{\Omega \Psi_j^2(a_1)}{M_1 \Phi_j(\omega)} \right|^2}{|\Delta|^2} \right] \sum_r \sum_s \frac{\Psi_r(a_2)\Psi_s(a_2)S_{rs}(\omega)}{M_2^2 \Phi_r^*(\omega)\Phi_s(\omega)}$$

$$-\mathrm{i}k_c\omega \left[\frac{\sum_r \frac{\Omega \Psi_r^2(a_2)}{M_2 \Phi_r(\omega)} + \left| \sum_r \frac{\Omega \Psi_r^2(a_2)}{M_2 \Phi_r(\omega)} \right|^2}{|\Delta|^2} \right] \sum_i \sum_j \frac{\Psi_i(a_1)\Psi_j(a_1)S_{ij}(\omega)}{M_1^2 \Phi_i^*(\omega)\Phi_j(\omega)}$$

$$+\frac{\gamma\omega^2}{|\Delta|^2} \left(\sum_j \frac{\Omega \Psi_j^2(a_1)}{M_1 \Phi_j(\omega)} \right)^* \sum_r \sum_s \frac{\Psi_r(a_2)\Psi_s(a_2)S_{rs}(\omega)}{M_2^2 \Phi_r^*(\omega)\Phi_s(\omega)}$$

$$+\frac{\gamma\omega^2}{|\Delta|^2} \left[-1 - \left(\sum_r \frac{\Omega \Psi_r^2(a_2)}{M_2 \Phi_r(\omega)} \right)^* \right] \sum_i \sum_j \frac{\Psi_i(a_1)\Psi_j(a_1)S_{ij}(\omega)}{M_1^2 \Phi_i^*(\omega)\Phi_j(\omega)}. \quad (3.17)$$

Now, when evaluating energy flows, integration is taken over even ranges in the frequency domain so that only the real and even part of the previous expression need be retained. After the necessary mathematical manipulations, and reversing the subscripts, the expression for the power transferred to subsystem 2 can then be recovered as

$$\Pi_{12}(\omega) = \frac{c_2\omega^2 |\Omega|^2}{M_2 M_1^2 |\Delta|^2} \sum_r \frac{\Psi_r^2(a_2)}{|\Phi_r(\omega)|^2} \sum_i \sum_j \frac{\Psi_i(a_1)\Psi_j(a_1)\mathrm{Re}\{\Phi_i(\omega)\Phi_j^*(\omega)\}S_{ij}(\omega)}{|\Phi_i(\omega)|^2 |\Phi_j(\omega)|^2}$$

$$-\frac{c_1\omega^2 |\Omega|^2}{M_1 M_2^2 |\Delta|^2} \sum_i \frac{\Psi_i^2(a_1)}{|\Phi_i(\omega)|^2} \sum_r \sum_s \frac{\Psi_r(a_2)\Psi_s(a_2)\mathrm{Re}\{\Phi_r(\omega)\Phi_s^*(\omega)\}S_{rs}(\omega)}{|\Phi_r(\omega)|^2 |\Phi_s(\omega)|^2}$$

$$-\frac{\gamma\omega^2}{M_2^2 |\Delta|^2} \sum_r \sum_s \frac{\Psi_r(a_2)\Psi_s(a_2)\mathrm{Re}\{\Phi_r(\omega)\Phi_s^*(\omega)\}S_{rs}(\omega)}{|\Phi_r(\omega)|^2 |\Phi_s(\omega)|^2}. \quad (3.18)$$

Notice that this is exactly as derived by Keane (1988) except that a complex coupling stiffness is used, i.e., Ω in place of k_c, and there is an additional term in $\gamma\omega^2$ (in fact, if the analysis presented by Keane (1988) is carried out starting with a complex coupling stiffness, this expression is exactly recovered).

(b) Input Power

The energy flowing into subsystem 1 from the external forcing is given by the expression

$$\Pi_{1_{in}}(t) = \int_1 \dot{y}_1(x_1,t)F_1(x_1,t)\mathrm{d}x_1. \quad (3.19)$$

The ensemble average of this equation is

$$\mathrm{E}[\Pi_{1_{in}}(t)] = \mathrm{E}[\Pi_{1_{in}}] = \int_1 \mathrm{E}[\dot{y}_1(x_1,t)F_1(x_1,t)]\mathrm{d}x_1. \quad (3.20)$$

By inserting the modal expansions for displacements, the last equation becomes

$$\mathrm{E}[\Pi_{1_{in}}(t)] = \sum_i \sum_j \mathrm{E}[\dot{W}_i(t) L_j(t)] \int_1 \frac{\Psi_i(x_1)\Psi_j(x_1)\rho_1(x_1)}{M_1} \mathrm{d}x_1. \tag{3.21}$$

Next, using the orthogonality property the last equation can be simplified to

$$\mathrm{E}[\Pi_{1_{in}}] = \sum_i \mathrm{E}[\dot{W}_i(t) L_i(t)]. \tag{3.22}$$

Taking spectral densities and Fourier transforms as before gives

$$\Pi_{1_{in}}(\omega) = \sum_i S_{\dot{W}_i i}(\omega) = -\mathrm{i}\omega \sum_i S_{W_i i}(\omega). \tag{3.23}$$

Here $S_{W_i i}$ is the cross-spectral density of the modal displacement and the modal driving force and is given by expression

$$S_{W_i i} = \lim_{T\to\infty} [W_i^*(\omega) L_i^*(\omega)] \frac{2\pi}{T}. \tag{3.24}$$

Substituting $W_i(\omega)$ from equation(2.33) gives

$$\Pi_{1_{in}}(\omega) = -\mathrm{i}\omega \sum_i \left(\lim_{T\to\infty} \left(\frac{\kappa_i^*(\omega) + L_i^*(\omega)}{M_1 \Phi_i^*(\omega)} L_i(\omega) \right) \frac{2\pi}{T} \right). \tag{3.25}$$

Then multiplying the various terms out gives

$$\Pi_{1_{in}}(\omega) = -\mathrm{i}\omega \sum_i \left(\frac{S_{ii}(\omega)}{M_1 \Phi_i^*(\omega)} + \frac{\Omega^* \Psi_i(a_1)}{\Delta^* M_1 \Phi_i^*(\omega)} \left(\sum_r \frac{\Psi_r(a_2) S_{ri}(\omega)}{M_2 \Phi_r^*(\omega)} - \sum_j \frac{\Psi_j(a_1) S_{ji}(\omega)}{M_1 \Phi_j^*(\omega)} \right) \right). \tag{3.26}$$

Assuming that the driving forces acting on the subsystems are uncorrelated as before gives

$$\Pi_{1_{in}}(\omega) = -\mathrm{i}\omega \sum_i \frac{S_{ii}(\omega)}{M_1 \Phi_i^*(\omega)} + \frac{-\mathrm{i}\omega\Omega^*}{M_1^2 \Delta^*} \sum_i \frac{\Psi_i(a_1)}{\Phi_i^*(\omega)} \sum_j \frac{\Psi_i(a_1) S_{ji}(\omega)}{\Phi_j^*(\omega)} \tag{3.27}$$

and taking only the real and even parts of the last expression leads to the desired expression for the input power as

$$\Pi_{1_{in}}(\omega) = \frac{\omega^2 c_1}{M_1} \sum_i \frac{S_{ii}(\omega)}{\mid \Phi_i(\omega) \mid^2} + \frac{\omega}{M_1^2} \mathrm{Im} \left\{ \sum_i \sum_j \frac{\Psi_i(a_1)\Psi_j(a_1)\Omega S_{ji}(\omega)}{\Phi_i(\omega)\Phi_j(\omega)\Delta} \right\}. \tag{3.28}$$

(c) Dissipated power

The power dissipated by damping in subsystem 1 is given by the expression

$$\Pi_{1_{diss}}(t) = \int_1 \dot{y}_1(x_1,t) r_1(x_1) \dot{y}_1(x_1,t) \mathrm{d}x_1. \tag{3.29}$$

The ensemble average is

$$\mathrm{E}[\Pi_{1_{diss}}(t)] = \mathrm{E}[\Pi_{1_{diss}}] = \int_1 \mathrm{E}[\dot{y}_1^2(x_1,t)] r_1(x_1) \mathrm{d}x_1 \tag{3.30}$$

so that by inserting modal expansions for the displacements the expression becomes

$$\mathrm{E}[\Pi_{1_{diss}}] = \sum_i \sum_j \mathrm{E}[\dot{W}_i(t)\dot{W}_j(t)] \int_1 \Psi_i(x_1)\Psi_j(x_1)r_1(x_1)\mathrm{d}x_1. \tag{3.31}$$

Now, since $r_1(x_1) = c_1\rho_1(x_1)$ and using the orthogonality property, this gives

$$\mathrm{E}[\Pi_{1_{diss}}] = c_1 M_1 \sum_i \mathrm{E}[\dot{W}_i(t)\dot{W}_i(t)]. \tag{3.32}$$

Taking the spectral densities and Fourier transforms leads to

$$\Pi_{1_{diss}}(\omega) = M_1 c_1 \sum_i S_{\dot{W}_i\dot{W}_i}(\omega) \quad \text{or} \quad \Pi_{1_{diss}}(\omega) = M_1 c_1 \omega^2 \sum_i S_{\dot{W}_i\dot{W}_i}(\omega) \tag{3.33,3.34}$$

where $S_{\dot{W}_i\dot{W}_i}$ is given by the expression

$$S_{\dot{W}_i\dot{W}_i} = \lim_{T\to\infty}[W_i^*(\omega)W_i(\omega)]\frac{2\pi}{T}. \tag{3.35}$$

Now substituting for $W_i(\omega)$ from equation (2.33) gives

$$\Pi_{1_{diss}}(\omega) = M_1 c_1 \omega^2 \sum_i \lim_{T\to\infty} \left\{ \frac{[L_i(\omega)+\kappa_i(\omega)]^*[L_i(\omega)+\kappa_i(\omega)]}{[M_1\Phi_i(\omega)]^*[M_1\Phi_i(\omega)]} \right\} \frac{2\pi}{T}. \tag{3.36}$$

So multiplying out the various terms leads to

$$\Pi_{1_{diss}}(\omega) = \frac{c_1\omega^2}{M_1} \sum_i \frac{S_{ii}(\omega)}{\mid \Phi_i(\omega) \mid^2}$$

$$-\frac{c_1\omega^2}{M_1^2} \sum_i \sum_j \frac{\Psi_i(a_1)\Psi_j(a_1)}{\mid \Phi_i(\omega) \mid^2} \left(\frac{\Omega^* S_{ji}(\omega)}{\Delta^*\Phi_j^*(\omega)} + \frac{\Omega^* S_{ij}(\omega)}{\Delta^*\Phi_j(\omega)} \right)$$

$$+\frac{c_1\omega^2}{M_1} \sum_i \frac{\mid \Omega \mid^2 \Psi_i^2(a_1)}{\mid \Delta \mid^2 \mid \Phi_i(\omega) \mid^2} \left(\sum_r \sum_s \frac{\Psi_r(a_2)\Psi_s(a_2)S_{rs}(\omega)}{M_2^2\Phi_r^*(\omega)\Phi_s(\omega)} + \sum_i \sum_j \frac{\Psi_i(a_1)\Psi_j(a_1)S_{ij}(\omega)}{M_1^2\Phi_i^*(\omega)\Phi_j(\omega)} \right). \tag{3.37}$$

Again taking only the real and even components finally gives

$$\Pi_{1_{diss}}(\omega) = \frac{c_1\omega^2}{M_1} \sum_i \frac{S_{ii}(\omega)}{\mid \Phi_i(\omega) \mid^2}$$

$$-2\frac{c_1\omega^2}{M_1^2} \sum_i \sum_j \frac{\Psi_i(a_1)\Psi_j(a_1)}{\mid \Phi_i(\omega) \mid^2} \mathrm{Re}\left\{ \frac{\Omega S_{ij}(\omega)}{\Delta\Phi_j(\omega)} \right\}$$

$$+\frac{c_1\omega^2}{M_1 \mid \Delta \mid^2} \sum_i \frac{\mid \Omega \mid^2 \Psi_i^2(a_1)}{\mid \Phi_i(\omega) \mid^2} \left\{ \sum_r \sum_s \frac{\Psi_r(a_2)\Psi_s(a_2)}{M_2^2} \mathrm{Re}\left\{ \frac{S_{rs}(\omega)}{\Phi_r^*(\omega)\Phi_s(\omega)} \right\} \right.$$

$$\left. + \sum_i \sum_j \frac{\Psi_i(a_1)\Psi_j(a_1)}{M_1^2} \mathrm{Re}\left\{ \frac{S_{ij}(\omega)}{\Phi_i^*(\omega)\Phi_j(\omega)} \right\} \right\}. \tag{3.38}$$

(d) Power dissipated within the coupling

The power dissipated within the coupling is given by the expression

$$\Pi_{dc}(t) = \gamma[\dot{y}_1(a_1,t) - \dot{y}_2(a_2,t)]^2 = -[\Pi_{12}(t) + \Pi_{21}(t)] \tag{3.39}$$

so that in the frequency domain

$$\Pi_{dc}(\omega) = -[\Pi_{12}(\omega) + \Pi_{21}(\omega)]$$

$$= \frac{\gamma\omega^2}{M_1^2 \mid \Delta \mid^2} \sum_i \sum_j \frac{\Psi_i(a_1)\Psi_j(a_1)\mathrm{Re}\{\Phi_i(\omega)\Phi_j^*(\omega)\}S_{ij}(\omega)}{\mid \Phi_i(\omega) \mid^2 \mid \Phi_j(\omega) \mid^2}$$

$$+ \frac{\gamma\omega^2}{M_2^2 \mid \Delta \mid^2} \sum_r \sum_s \frac{\Psi_r(a_2)\Psi_s(a_2)\mathrm{Re}\{\Phi_r(\omega)\Phi_s^*(\omega)\}S_{rs}(\omega)}{\mid \Phi_r(\omega) \mid^2 \mid \Phi_s(\omega) \mid^2} \tag{3.40}$$

(e) Energy levels

Here the energy level of a subsystem is taken to be twice its kinetic energy, which for subsystem 1 is given by the expression

$$KE_1(t) = \frac{1}{2}\int_1 \dot{y}_1^2(x_1,t)\rho_1(x_1)\mathrm{d}x_1 \tag{3.41}$$

so that

$$E_1(t) = \int_1 \dot{y}_1^2(x_1,t)\rho_1(x_1)\mathrm{d}x_1 = \Pi_{1_{diss}}(t)/c_1 \tag{3.42}$$

so that, in the frequency domain, the expression for the energy level is given by

$$E_1(\omega) = \frac{\omega^2}{M_1} \sum_i \frac{S_{ii}(\omega)}{\mid \Phi_i(\omega) \mid^2}$$

$$-2\frac{\omega^2}{M_1^2} \sum_i \sum_j \frac{\Psi_i(a_i)\Psi_j(a_i)}{\mid \Phi_i(\omega) \mid^2} \mathrm{Re}\left\{\frac{\Omega S_{ij}(\omega)}{\Delta\Phi_j(\omega)}\right\}$$

$$+\frac{\omega^2}{M_1 \mid \Delta \mid^2} \sum_i \frac{\mid \Omega \mid^2 \Psi_i^2(a_1)}{\mid \Phi_i(\omega) \mid^2} \left(\sum_r \sum_s \frac{\Psi_r(a_2)\Psi_s(a_2)}{M_2^2} \mathrm{Re}\left\{\frac{S_{rs}(\omega)}{\Phi_r^*(\omega)\Phi_s(\omega)}\right\} \right.$$

$$\left. + \sum_i \sum_j \frac{\Psi_i(a_1)\Psi_j(a_1)}{M_1^2} \mathrm{Re}\left\{\frac{S_{ij}(\omega)}{\Phi_i^*(\omega)\Phi_j(\omega)}\right\} \right) \tag{3.43}$$

4. Two forcing models

The relation between the modal spectrum $S_{ij}(\omega)$ and the subsystem forcing spectrum is simply

$$\int\int S_{F_1F_1}(\omega, x_i, x_j)\Psi_i(x_i)\Psi_j(x_j)\mathrm{d}x_i\mathrm{d}x_j = S_{ij}(\omega). \tag{4.1}$$

It is convenient to consider forcing models that allow the separation of frequency and spatial variables in the expression of modal spectra. The first model considered is point driving.

(a) Point driving

When the driving forces are applied at a single point $x_1 = b_1$ on subsystem 1 and at $x_2 = b_2$ on subsystem 2, the expression for the forcing spectrum of subsystem 1 becomes

$$S_{F_1F_1}(\omega, x_1, x_1) = S_{F_1F_1}(\omega)\delta(x_1 - b_1)\delta(x_1 - b_1). \tag{4.2}$$

Substituting in equation (4.1), the expression for the modal spectra of subsystem 1 becomes

$$S_{ij}(\omega) = S_{F_1F_1}(\omega)\Psi_i(b_1)\Psi_j(b_1). \tag{4.3}$$

The expression for the energy flowing into subsystem 2 for this model of forcing is then

$$\begin{aligned}\Pi_{12}(\omega) = {} & \frac{c_2\omega^2 \mid \Omega \mid^2}{M_2M_1^2 \mid \Delta \mid^2} \sum_r \frac{\Psi_r^2(a_2)}{\mid \Phi_r(\omega) \mid^2} \left| \sum_i \frac{\Psi_i(a_1)\Psi_i(b_1)}{\Phi_i(\omega)} \right|^2 S_{F_1F_1}(\omega) \\ & - \frac{c_1\omega^2 \mid \Omega \mid^2}{M_1M_2^2 \mid \Delta \mid^2} \sum_i \frac{\Psi_i^2(a_1)}{\mid \Phi_i(\omega) \mid^2} \left| \sum_r \frac{\Psi_r(a_2)\Psi_r(b_2)}{\Phi_r(\omega)} \right|^2 S_{F_2F_2}(\omega) \\ & - \frac{\gamma\omega^2}{M_2^2 \mid \Delta \mid^2} \left| \sum_r \frac{\Psi_r(a_2)\Psi_r(b_2)}{\Phi_r(\omega)} \right|^2 S_{F_2F_2}(\omega)\end{aligned} \tag{4.4}$$

that for the power dissipated in the coupling becomes

$$\begin{aligned}\Pi_{dc}(\omega) = {} & \frac{\gamma\omega^2}{M_1^2 \mid \Delta \mid^2} \left| \sum_i \frac{\Psi_i(a_1)\Psi_i(b_1)}{\Phi_i(\omega)} \right|^2 S_{F_1F_1}(\omega) \\ & + \frac{\gamma\omega^2}{M_2^2 \mid \Delta \mid^2} \left| \sum_r \frac{\Psi_r(a_2)\Psi_i(b_2)}{\Phi_r(\omega)} \right|^2 S_{F_2F_2}(\omega)\end{aligned} \tag{4.5}$$

and the expression for the power input into subsystem 1 is

$$\Pi_{1_{in}}(\omega) = \frac{\omega^2 c_1}{M_1} \sum_i \frac{\Psi_i^2(a_1)}{\mid \Phi_i(\omega) \mid^2} S_{F_1F_1}(\omega) + \frac{\omega}{M_1^2} \mathrm{Im} \left\{ \frac{\Omega}{\Delta} \left(\sum_i \frac{\Psi_i(a_1)\Psi_i(b_1)}{\Phi_i(\omega)} \right)^2 \right\} S_{F_1F_1}(\omega). \tag{4.6}$$

(b) Rain-on-the-roof model

In this model of forcing, the modal components of the driving forces are incoherent and this implies that

$$S_{ij}(\omega) = \int\!\!\int S_{F_1F_1}(\omega, x_i, x_j)\Psi_i(x_i)\Psi_j(x_j)\mathrm{d}x_i\mathrm{d}x_j = S_{ij}(\omega)\delta_{ij}. \tag{4.7}$$

In this case, the expression for the forcing spectrum for subsystem 1 is

$$S_{F_1F_1}(\omega, x_1, x_1') = S_{F_1F_1}(\omega)\delta(x_1 - x_1')\frac{\rho_1}{M_1}. \tag{4.8}$$

The expression for the energy flowing into subsystem 2 for this model of forcing is

$$\Pi_{12}(\omega) = \frac{c_2\omega^2 \mid \Omega \mid^2}{M_2 M_1^2 \mid \Delta \mid^2} \sum_r \frac{\Psi_r^2(a_2)}{\mid \Phi_r(\omega) \mid^2} \sum_i \frac{\Psi_i^2(a_1)}{\mid \Phi_i(\omega) \mid^2} S_{F_1F_1}(\omega)$$
$$- \frac{c_1\omega^2 \mid \Omega \mid^2}{M_2 M_2^2 \mid \Delta \mid^2} \sum_i \frac{\Psi_i^2(a_1)}{\mid \Phi_i(\omega) \mid^2} \sum_r \frac{\Psi_r^2(a_2)}{\mid \Phi_r(\omega) \mid^2} S_{F_2F_2}(\omega)$$
$$- \frac{\gamma\omega^2}{M_2^2 \mid \Delta \mid^2} \sum_r \frac{\Psi_r^2(a_2)}{\mid \Phi_r(\omega) \mid^2} S_{F_2F_2}(\omega) \tag{4.9}$$

and that for the power dissipated in the coupling becomes

$$\Pi_{dc}(\omega) = \frac{\gamma\omega^2}{M_1^2 \mid \Delta \mid^2} \sum_i \frac{\Psi_i^2(a_1)}{\mid \Phi_i(\omega) \mid^2} S_{F_1F_1}(\omega) + \frac{[\gamma\omega^2]}{M_2^2 \mid \Delta \mid^2} \sum_r \frac{\Psi_r^2(a_2)}{\mid \Phi_r(\omega) \mid^2} S_{F_2F_2}(\omega). \tag{4.10}$$

Finally, the expression for the power input into subsystem 1 becomes

$$\Pi_{1_{in}}(\omega) = \frac{\omega^2 c_1}{M_1} \sum_i \frac{1}{\mid \Phi_i(\omega) \mid^2} S_{F_1F_1}(\omega) + \frac{\omega}{M_1^2} \mathrm{Im} \left\{ \frac{\Omega}{\Delta} \left(\sum_i \frac{\Psi_i(a_1)}{\Phi_i(\omega)} \right)^2 \right\} S_{F_1F_1}(\omega). \tag{4.11}$$

(c) Power Receptances

The various expressions for the input power, coupling power and power dissipated within the coupling can be written in terms of power receptances for the two cases of forcing as follows:

$$\Pi_{1_{in}}(\omega) = H_{11}(\omega) S_{F_1F_1}(\omega), \tag{4.12}$$

$$\Pi_{2_{in}}(\omega) = H_{22}(\omega) S_{F_2F_2}(\omega), \tag{4.13}$$

$$\Pi_{dc}(\omega) = H_{dc_1}(\omega) S_{F_1F_1}(\omega) + H_{dc_2}(\omega) S_{F_2F_2}(\omega), \tag{4.14}$$

$$\Pi_{12}(\omega) = H_{12}(\omega) S_{F_1F_1}(\omega) - H_{21}(\omega) S_{F_2F_2}(\omega) - H_{dc_2}(\omega) S_{F_2F_2}(\omega), \tag{4.15}$$

$$\Pi_{21}(\omega) = H_{21}(\omega) S_{F_2F_2}(\omega) - H_{12}(\omega) S_{F_1F_1}(\omega) - H_{dc_1}(\omega) S_{F_1F_1}(\omega), \tag{4.16}$$

where the expressions for the various receptances are

(i) point driving

$$H_{11}(\omega) = \frac{\omega^2 c_1}{M_1} \sum_i \frac{\Psi_i^2(a_1)}{\mid \Phi_i(\omega) \mid^2} + \frac{\omega}{M_1^2} \mathrm{Im} \left\{ \frac{\Omega}{\Delta} \left(\sum_i \frac{\Psi_i(a_1)\Psi_i(b_1)}{\Phi_i(\omega)} \right)^2 \right\},$$

$$H_{22}(\omega) = \frac{\omega^2 c_2}{M_2} \sum_r \frac{\Psi_r^2(a_2)}{\mid \Phi_r(\omega) \mid^2} + \frac{\omega}{M_2^2} \mathrm{Im} \left\{ \frac{\Omega}{\Delta} \left(\sum_r \frac{\Psi_r(a_2)\Psi_r(b_2)}{\Phi_r(\omega)} \right)^2 \right\},$$

$$H_{12}(\omega) = \frac{c_2\omega^2 \mid \Omega \mid^2}{M_2 M_1^2 \mid \Delta \mid^2} \sum_r \frac{\Psi_r^2(a_2)}{\mid \Phi_r(\omega) \mid^2} \left| \sum_i \frac{\Psi_i(a_1)\Psi_i(b_1)}{\Phi_i(\omega)} \right|^2,$$

$$H_{21}(\omega) = \frac{c_1\omega^2 \mid \Omega \mid^2}{M_1 M_2^2 \mid \Delta \mid^2} \sum_i \frac{\Psi_i^2(a_1)}{\mid \Phi_i(\omega) \mid^2} \left| \sum_r \frac{\Psi_r(a_2)\Psi_r(b_2)}{\Phi_r(\omega)} \right|^2,$$

Table 1. *Parameters used in the examples*

Parameter	Rod 1	Rod 2	Units
Mass density(ρ)	4.156	4.156	kg/m
Length (l)	5.182	4.328	m
Rigidity (EA)	17.85	17.85	MN
Coupling Position (x/l)	0.1176	0.7042	–
Damping strength (c)	88.95	106.49	s^{-1}

$$H_{dc_1}(\omega) = \frac{\gamma\omega^2}{M_1^2 \mid \Delta \mid^2} \sum_i \frac{\Psi_i^2(a_1)}{\mid \Phi_i(\omega) \mid^2},$$

$$H_{dc_2}(\omega) = \frac{\gamma\omega^2}{M_2^2 \mid \Delta \mid^2} \sum_r \frac{\Psi_r^2(a_2)}{\mid \Phi_r(\omega) \mid^2};$$

(ii) rain-on-the-roof driving

$$H_{11}(\omega) = \frac{\omega^2 c_1}{M_1} \sum_i \frac{1}{\mid \Phi_i(\omega) \mid^2} + \frac{\omega}{M_1^2} \mathrm{Im} \left\{ \frac{\Omega}{\Delta} \left(\sum_i \frac{\Psi_i(a_1)}{\Phi_i(\omega)} \right)^2 \right\},$$

$$H_{22}(\omega) = \frac{\omega^2 c_2}{M_2} \sum_r \frac{1}{\mid \Phi_r(\omega) \mid^2} + \frac{\omega}{M_2^2} \mathrm{Im} \left\{ \frac{\Omega}{\Delta} \left(\sum_r \frac{\Psi_r(a_2)}{\Phi_r(\omega)} \right)^2 \right\},$$

$$H_{12}(\omega) = \frac{c_2\omega^2 \mid \Omega \mid^2}{M_2 M_1^2 \mid \Delta \mid^2} \sum_r \frac{\Psi_r^2(a_2)}{\mid \Phi_r(\omega) \mid^2} \sum_i \frac{\Psi_i^2(a_1)}{\mid \Phi_i(\omega) \mid^2},$$

$$H_{21}(\omega) = \frac{c_1\omega^2 \mid \Omega \mid^2}{M_1 M_2^2 \mid \Delta \mid^2} \sum_i \frac{\Psi_i^2(a_1)}{\mid \Phi_i(\omega) \mid^2} \sum_r \frac{\Psi_r^2(a_2)}{\mid \Phi_r(\omega) \mid^2},$$

$$H_{dc_1}(\omega) = \frac{\gamma\omega^2}{M_1^2 \mid \Delta \mid^2} \sum_i \frac{\Psi_i^2(a_1)}{\mid \Phi_i(\omega) \mid^2},$$

$$H_{dc_2}(\omega) = \frac{\gamma\omega^2}{M_2^2 \mid \Delta \mid^2} \sum_r \frac{\Psi_r^2(a_2)}{\mid \Phi_r(\omega) \mid^2}.$$

5. Variations in the coupling parameters

In the following examples rod 1 is loaded by unit forcing, so that the energy flow into subsystem 2 is equal to $H_{12}(\omega)$ and the power dissipated within the coupling is equal to $H_{dc_1}(\omega)$. The energy flow from rod 1 into the coupling is simply $H_{12}(\omega) + H_{dc_1}(\omega)$, whereas that input into rod 1 is $H_{11}(\omega)$. The properties of these various receptances are illustrated for the case of 'rain-on-the-roof' forcing and light damping. To aid comparison of the various results the parameters values adopted here are the same as those used in several previous studies (Remington and Manning, 1975; Keane and Price, 1987, 1991).

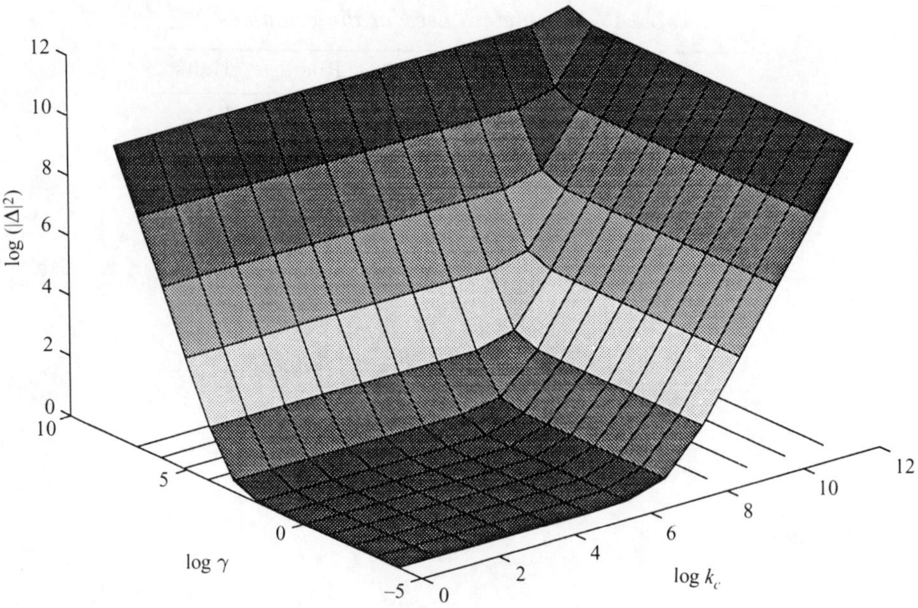

Figure 2. Variation in $\log(|\Delta|^2)$ with γ and k_c at $\omega = 10\,000$ rad/s.

(a) Coupling Strength

In the previous equations Δ may be written as

$$\Delta = 1 + \Omega \mathrm{Z} \tag{5.1}$$

where Z is a complex quantity which is independent of the coupling parameters γ and k_c. It is given by the expression

$$\mathrm{Z} = \sum_i \frac{\Psi_i^2(a_1)}{M_1 \Phi_i(\omega)} + \sum_r \frac{\Psi_r^2(a_2)}{M_2 \Phi_r(\omega)}. \tag{5.2}$$

Weak coupling may be defined by the criterion

$$|\,\Delta\,|^2 \approx 1$$

which requires that $|\,\Omega\,|$ is small. Then, as the coupling strength $|\,\Omega\,|$ increases, $|\,\Delta\,|^2$ also increases until the coupling becomes infinitely strong as $|\,\Delta\,|^2$ approaches infinity. The variation of $|\,\Delta\,|^2$ with the coupling parameters γ and k_c is illustrated in figure 2.

(b) Energy flow into rod 2

The cross-receptance $H_{12}(\omega)$ for the two cases of forcing discussed above can be written as

$$H_{12}(\omega) = P_{12} \frac{|\,\Omega\,|^2}{|\,\Delta\,|^2} \tag{5.3}$$

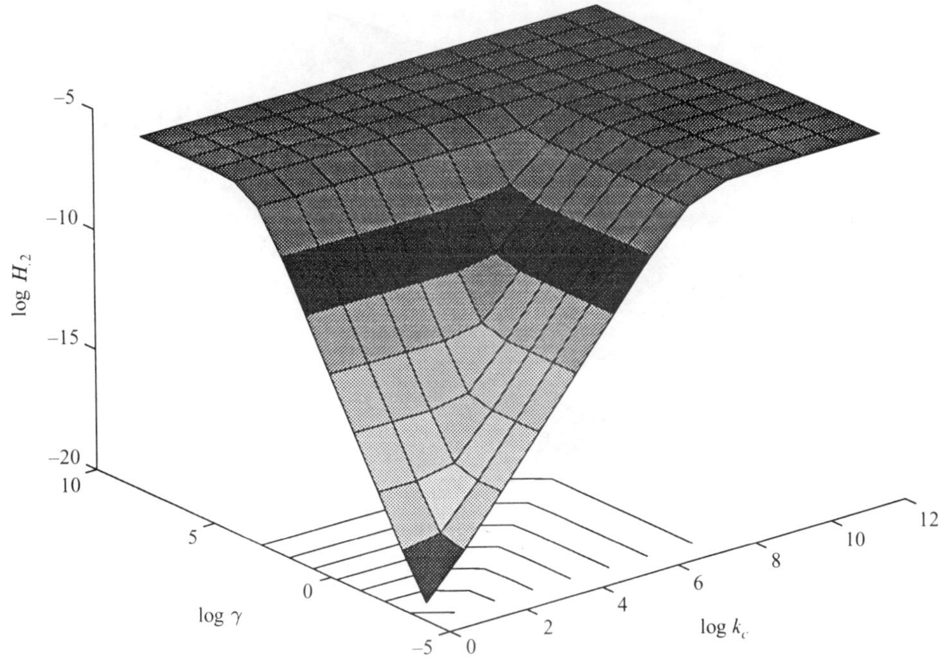

Figure 3. Variation in $\log(H_{12})$ with γ and k_c at $\omega = 10\,000$ rad/s.

where P_{12} contains all the terms that are independent of the coupling parameters. For the case of weak coupling it is clear that this reduces to

$$H_{12}(\omega) = P_{12} | \Omega |^2 \tag{5.4}$$

which implies that for weak coupling and a given magnitude of $| \Omega |^2$, the energy flow into rod 2 is independent of the ratio $\frac{\gamma\omega}{k_c}$. Then, as the coupling strength increases, the energy flow into rod 2 increases until $k_c \to \infty$ or $\gamma \to \infty$, when the energy flow reaches a constant level given by

$$\lim_{\gamma\to\infty} H_{12}(\omega) = \lim_{k_c\to\infty} H_{12}(\omega) = \frac{P_{12}}{| \mathrm{Z} |^2}. \tag{5.5}$$

This is illustrated in figure 3, which shows the variation of the cross-receptance $H_{12}(\omega)$ with respect to the two coupling parameters for a constant value of driving frequency ω. The variations of $H_{12}(\omega)$ with respect to the driving frequency ω and the coupling damping γ are illustrated in figure 4 for a fixed value of k_c. For $\gamma = 0$, the results are identical to those illustrated in figure 10 of Keane and Price's work (1991), which is plotted for conservative coupling. Figure 4 clearly shows peaks at the natural frequencies of both rods and marked dips between these frequencies. However, as the coupling damping strength increases, although the curves show peaks and dips at the same frequencies, the magnitude of $H_{12}(\omega)$ increases, as expected. This figure is also consistent with figure 8 of Chen and Soong's work (1991). Similarly, the variations of $H_{12}(\omega)$ with respect to driving frequency ω and spring stiffness k_c, for a fixed value of γ, is illustrated in figure 5. This is very similar to figure 4, which is as expected, since the relation between $H_{12}(\omega)$ and

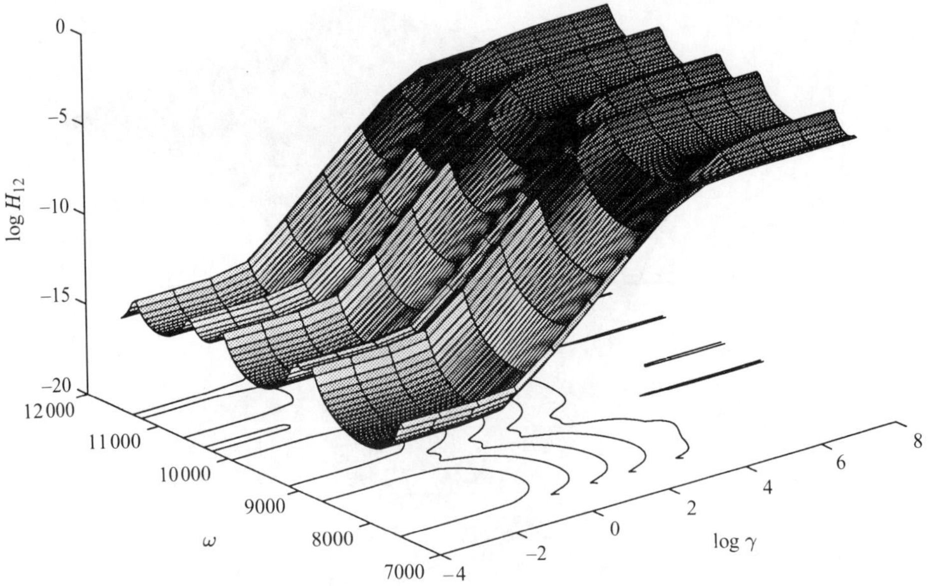

Figure 4. Variation in $\log(H_{12})$ with γ and ω for $k_c = 10\,000$ n/m.

k_c is identical to that between $H_{12}(\omega)$ and $\gamma\omega$. Both figures show that as the coupling becomes infinitely strong energy flow becomes independent of the coupling strength.

(c) Power dissipated within the coupling

For the case of conservative coupling, when $\gamma = 0$, there is no power dissipation within the coupling. However, even for a nonconservative coupling, when either $k_c \to \infty$ or $\gamma \to \infty$ the power dissipated within the coupling is also zero. This means that there must be certain values of γ and k_c for which the dissipated power is maximized. Maximum power is found to be dissipated within the coupling when

$$\gamma = \sqrt{\frac{1 + k_c^2 \mid \mathrm{Z} \mid^2 + 2k_c \mathrm{Re}\{\mathrm{Z}\}}{\omega^2 \mid \mathrm{Z} \mid^2}}. \tag{5.6}$$

Figure 6 clearly shows this behaviour, with the power dissipated within the coupling staying at relatively low levels except for a specific range of values of γ and k_c. Moreover, this behaviour is sensibly independent of k_c until k_c reaches sufficiently high levels that the damping element becomes rigidly blocked and the dissipated power then falls to low levels. It is straightforward to show that when the coupling is weak the power dissipated within the coupling is independent of k_c and takes the form

$$H_{dc_1}(\omega) = C\,\gamma, \tag{5.7}$$

where C is independent only on the subsystem parameters, see figure 7. The variation of $H_{dc_1}(\omega)$ with driving frequency ω and damping stiffness γ for a specific value of k_c is further illustrated in figure 8 and that for fixed γ with varying ω and k_c in figure 9. Both of these figures show peaks at the natural frequencies of the directly-driven subsystem and dips between these natural frequencies. Notice that figure 8 shows a maximum for

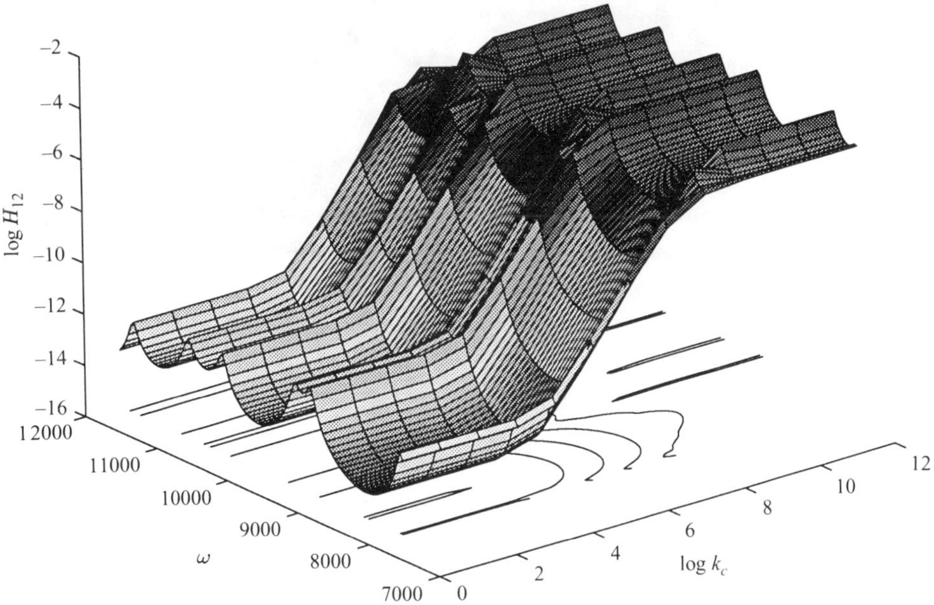

Figure 5. Variation in $\log(H_{12})$ with k_c and ω for $\gamma = 1$ ns/m.

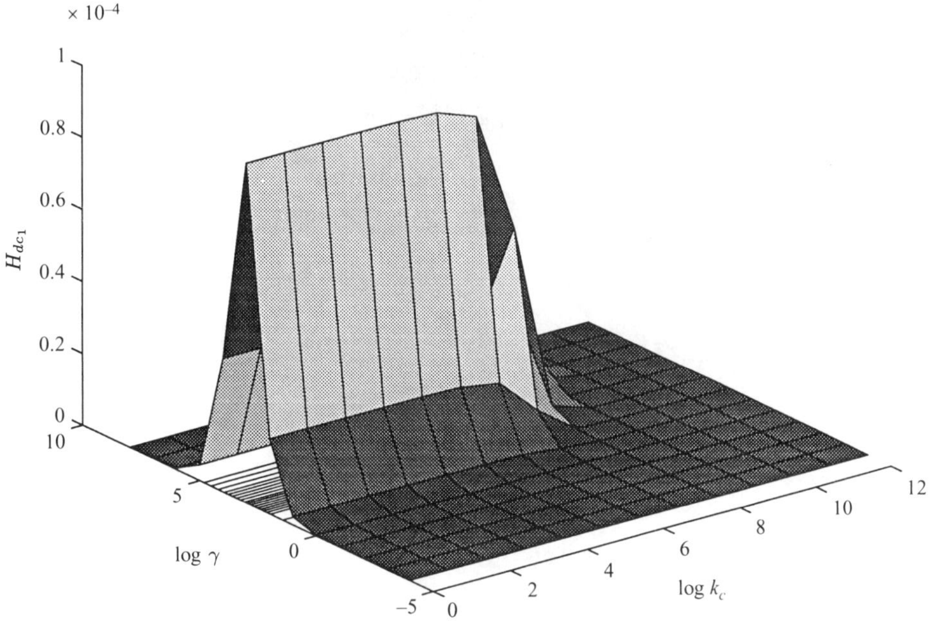

Figure 6. Variation in (H_{dc_1}) with γ and k_c at $\omega = 10\,000$ rad/s.

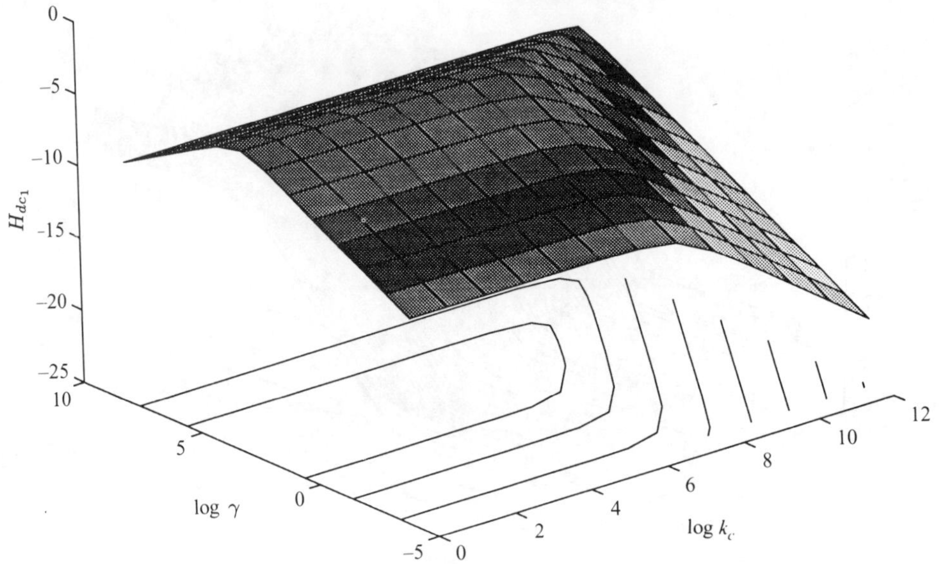

Figure 7. Variation in $\log(H_{dc_1})$ with γ and k_c at $\omega = 10\,000$ rad/s.

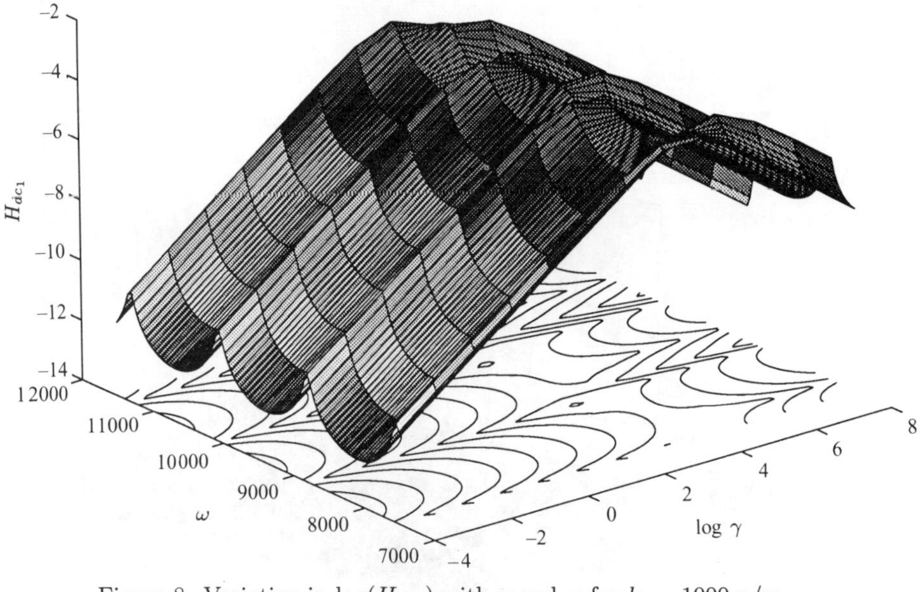

Figure 8. Variation in $\log(H_{dc_1})$ with γ and ω for $k_c = 1000$ n/m.

given ω at the same values of γ, while there are no maxima in figure 9 for variations in k_c, i.e., as per figure 6.

(d) Energy dissipated within the coupling compared with that transferred to rod 2

The energy that leaves rod 1 may be divided into two parts: some of it is dissipated within the coupling and the remainder is transferred to rod 2. It is of interest to see how

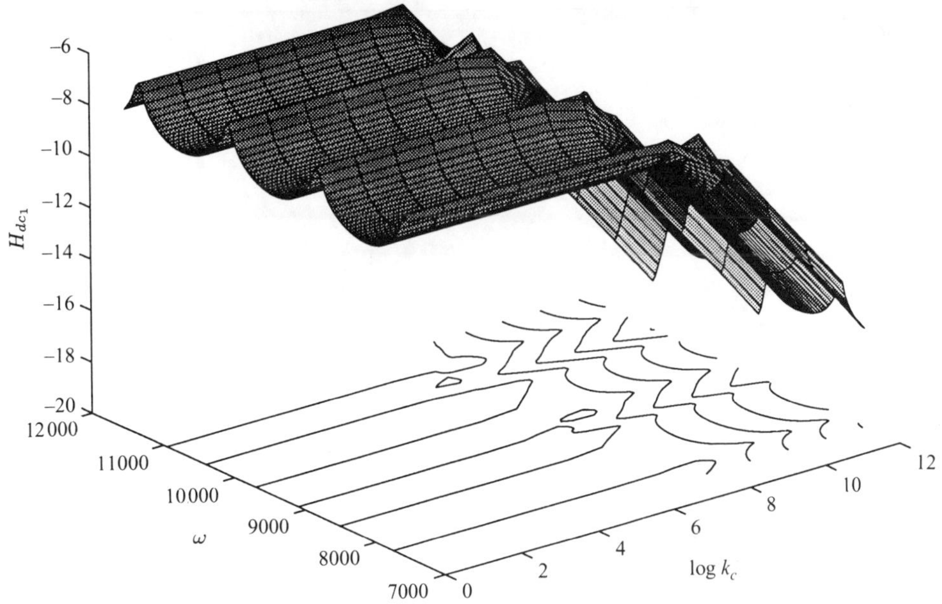

Figure 9. Variation in $\log(H_{dc_1})$ with k_c and ω for $\gamma = 1$ ns/m.

the ratio of these two quantities is affected by changes in the coupling parameters. This ratio is given by the expression

$$\frac{H_{dc_1}(\omega)}{H_{12}(\omega)} = \frac{\gamma B}{\mid \Omega \mid^2} = R \quad \text{where} \quad B = \frac{M_2}{c_2 \sum_r \dfrac{\Psi_r^2(a_2)}{\mid \Phi_r(\omega) \mid^2}}. \tag{5.8,5.9}$$

It is clear that the ratio is dependent only on the second rod and the coupling parameters, i.e., the ratio is not affected by any changes in the first rod's parameters. As the spring strength k_c increases, more power is transferred to rod 2 and less power is dissipated within the coupling, so that the ratio R decreases. However, as the coupling damping increases, R increases until it reaches a maximum value, when $\gamma = k_c/\omega$, after which it begins to fall again, see figure 10. This figure also shows the contour for which $\mid \Delta \mid^2 = 2$, i.e., where the strength of the coupling is transitional between weak and strong coupling, and it is clear that large values of R (i.e., those for which the losses in the coupling are a significant fraction of the coupling power) arise both when the coupling is weak and as it becomes strong.

A comparison of figure 6, which shows the absolute magnitude of the energy absorbed in the coupling, and figure 10 (which further shows the locus of maximum $H_{dc_1}(\omega)$ taken from figure 6) thus reveals that for moderate values of γ and k_c, the power dissipated in the coupling is usually much larger than that transmitted and that when it is not, this mostly arises because the coupling is essentially 'short-circuited' by either a very stiff spring or a virtually rigid damper. It is also clear from the elliptical nature of the K contours in figure 10, that it is possible to have low energy absorption in the damper if it is made too weak. Thus, there is a strictly bounded region of values of the two coupling parameters where the damper is absorbing both a significant proportion of the

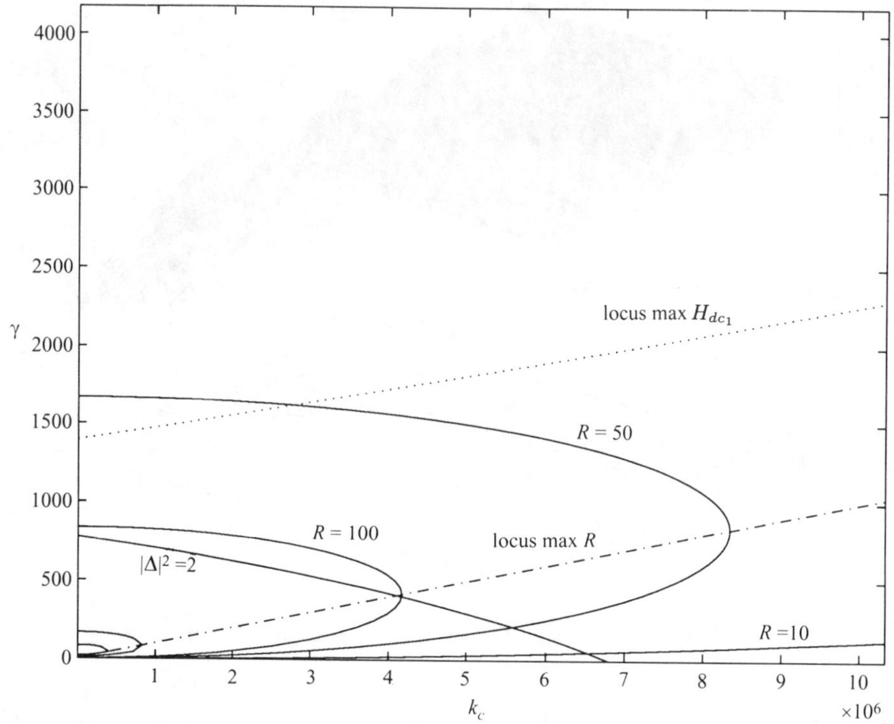

Figure 10. Variation in R with γ and k_c at $\omega = 10\,000$ rad/s, also showing contour of $|\Delta|^2 = 2$ and locus of manimum H_{dc_1}.

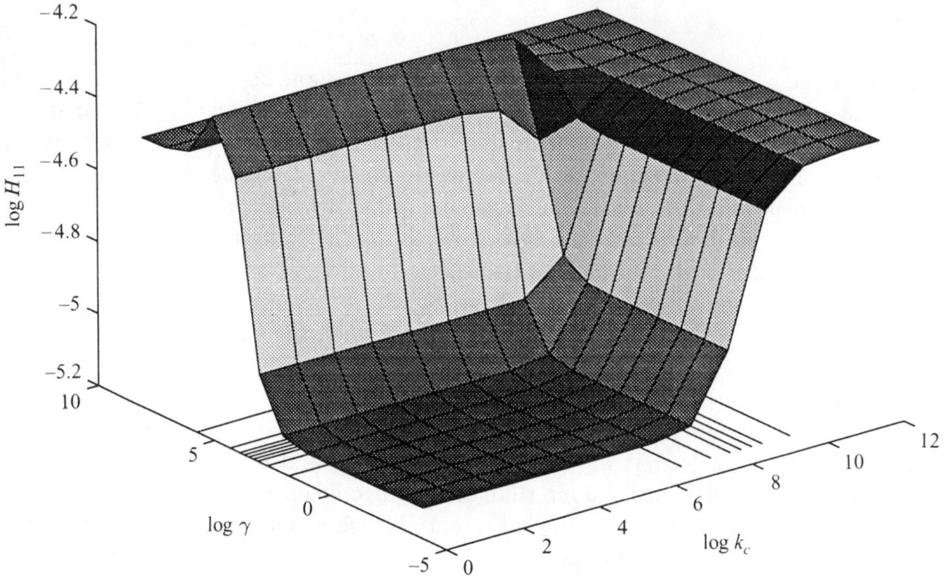

Figure 11. Variation in $\log(H_{11})$ with γ and k_c for $\omega = 8000$ rad/s.

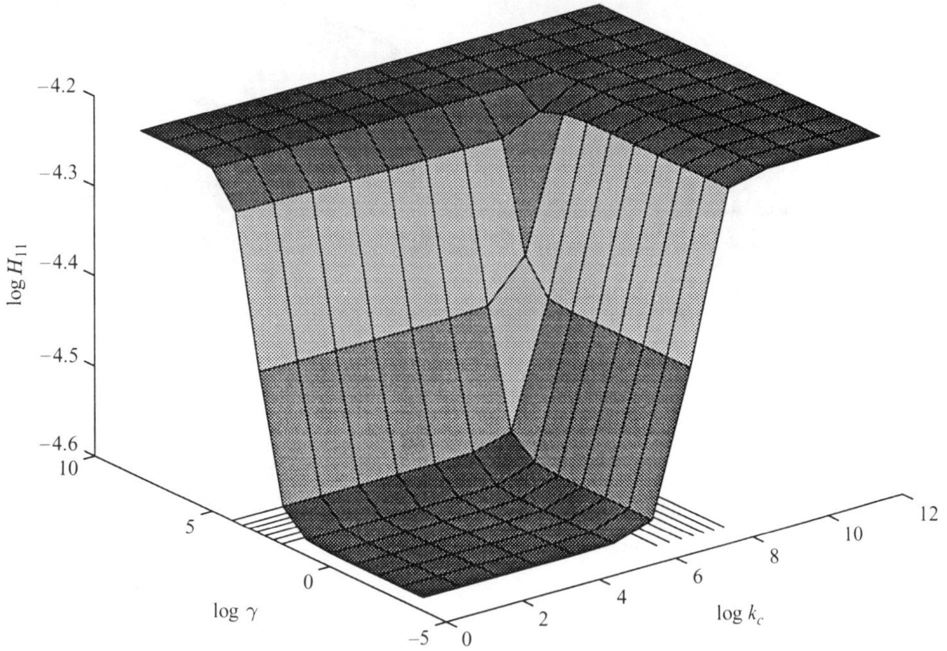

Figure 12. Variation in $\log(H_{11})$ with γ and k_c for $\omega = 9000$ rad/s.

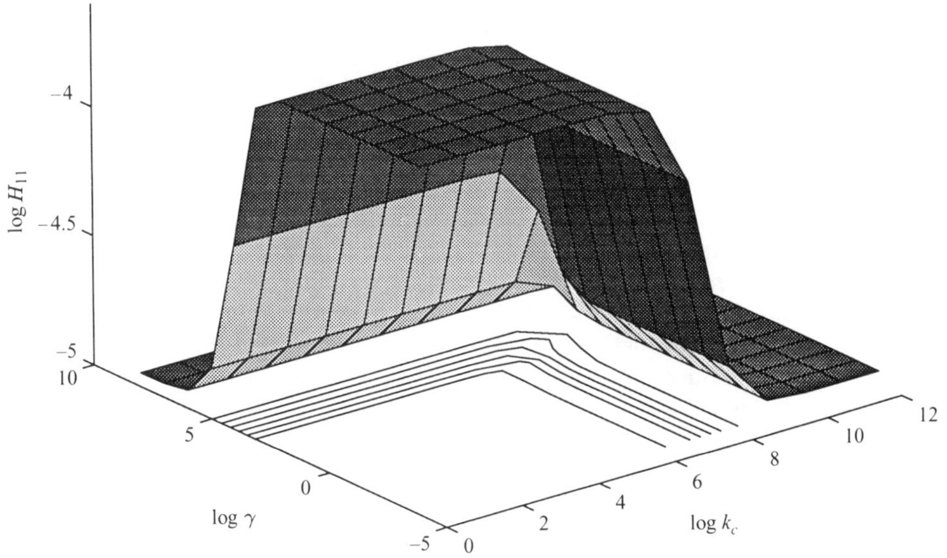

Figure 13. Variation in $\log(H_{11})$ with γ and k_c for $\omega = 10\,000$ rad/s.

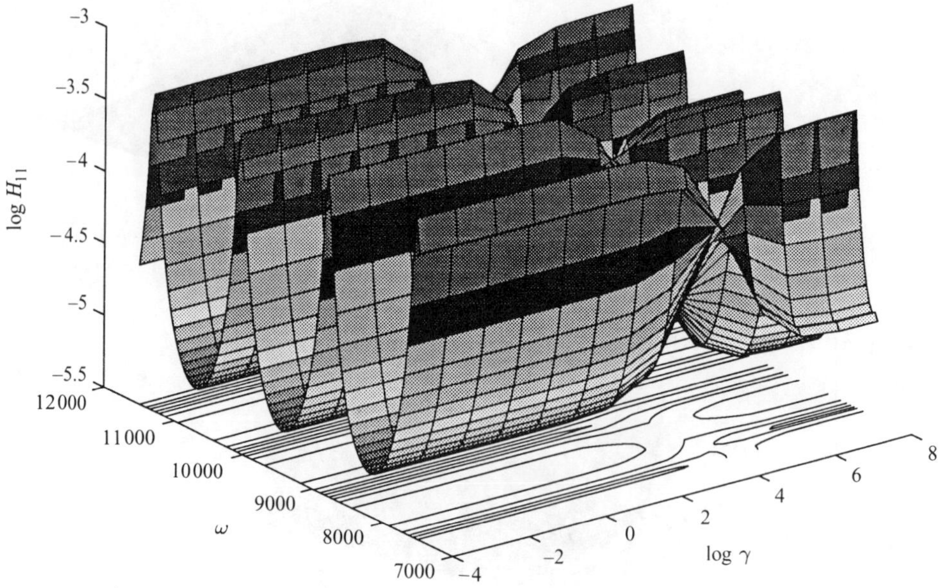

Figure 14. Variation in $\log(H_{11})$ with γ and ω for $k_c = 1000$ n/m.

transmitted energy and an amount which is also a significant fraction of that injected into the overall system. This kind of behaviour is well known to those who design shock mounts for sensitive equipment, where the damping in the mount must not be either too strong or too weak. In either case little energy is absorbed, in the first because the damper is hardly deflected and in the second because, although it deflects, it is too weak to have much effect.

(e) Input power

It can be seen from equation (4.11) that, for weak coupling, the first term in the expression dominates so that the input power is constant for small changes in k_c or γ. Since the second term in the expression may have positive or negative values depending on ω, variation of the input power for larger changes in the coupling parameters shows a variety of behaviours depending on the value of ω chosen, see figures 11–13. However, all these figures show two sensibly constant power levels and a transition from one to the other over roughly the same range of values of the coupling parameters, i.e., those that separate weak and strong coupling. The variation of $H_{11}(\omega)$ with driving frequency and γ for a constant value of k_c is illustrated in figure 14, and that with the driving frequency and k_c for constant value of γ in figure 15. Both figures show peaks at the natural frequencies of the directly-driven subsystem at low coupling strengths and shifts in these as the coupling becomes strong, as expected.

6. Average energy flows and subsystem energies

In the previous sections exact expressions for the various energy flows have been derived. Assuming that deterministic knowledge of the subsystems is not available, the subsystems characteristics can only be described probabilistically. All the previous equations are then expressed as ensemble averages, which are taken across a supposedly infinite set

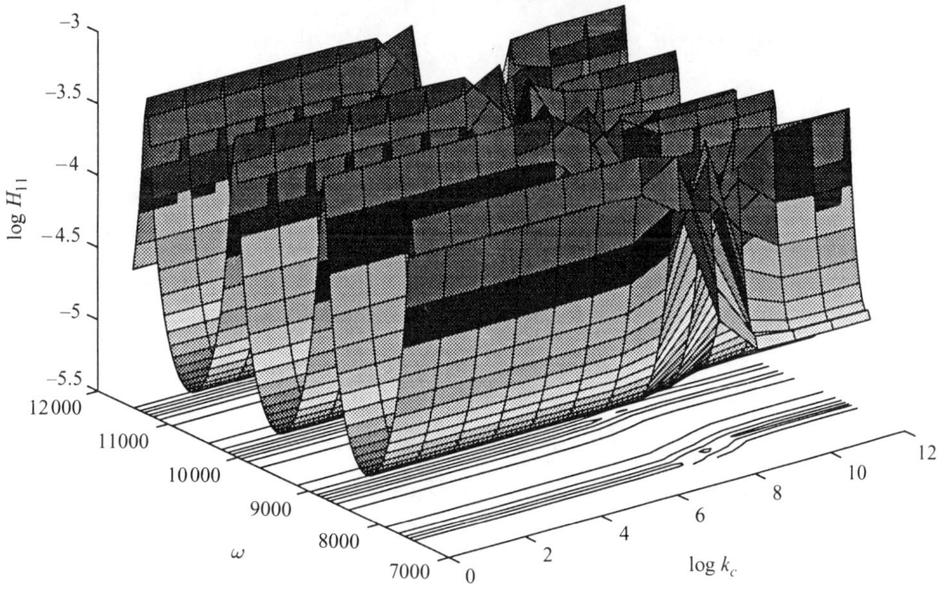

Figure 15. Variation in $\log(H_{11})$ with k_c and ω for $\gamma = 1$ ns/m.

of grossly similar systems, in which the individual members differ in some unpredictable detail. Ensemble averages are denoted by E[ ] and are functions of the driving frequency ω. Thus,

$$\mathrm{E}[\Pi_{1_{in}}(\omega)] = \mathrm{E}[H_{11}(\omega)]S_{F_1}S_{F_1}(\omega), \quad \mathrm{E}[\Pi_{1_{diss}}(\omega)] = c_1\mathrm{E}[E_1(\omega)] \tag{6.1,6.2}$$

and

$$\mathrm{E}[\Pi'_{12}(\omega)] = \mathrm{E}[H_{12}(\omega)]S_{F_1}S_{F_1}(\omega) + \mathrm{E}[H_{dc_1}(\omega)]S_{F_1}S_{F_1}(\omega) - \mathrm{E}[H_{21}(\omega)]S_{F_2}S_{F_2}(\omega). \tag{6.3}$$

Also

$$\mathrm{E}[\Pi_{dc}(\omega)] = \mathrm{E}[H_{dc_1}(\omega)]S_{F_1}S_{F_1}(\omega) + \mathrm{E}[H_{dc_2}(\omega)]S_{F_2}S_{F_2}(\omega). \tag{6.4}$$

The energy balance equations can be written as follows:

$$\mathrm{E}[\Pi_{1_{in}}(\omega)] - \mathrm{E}[\Pi_{1_{diss}}(\omega)] - \mathrm{E}[\Pi'_{12}(\omega)] = 0, \tag{6.5}$$

$$\mathrm{E}[\Pi_{2_{in}}(\omega)] - \mathrm{E}[\Pi_{2_{diss}}(\omega)] - \mathrm{E}[\Pi'_{21}(\omega)] = 0 \quad \text{and} \quad \mathrm{E}[\Pi'_{12}(\omega)] + \mathrm{E}[\Pi'_{21}(\omega)] = \mathrm{E}[\Pi_{dc}(\omega)]. \tag{6.6,6.7}$$

Rearranging these leads to

$$\mathrm{E}[\Pi'_{12}(\omega)] = \alpha_1\mathrm{E}[E_1(\omega)] - \alpha_2\mathrm{E}[E_2(\omega)] + \beta_1\mathrm{E}[E_1(\omega)] \tag{6.8}$$

and

$$\mathrm{E}[\Pi'_{21}(\omega)] = \alpha_2\mathrm{E}[E_2(\omega)] - \alpha_1\mathrm{E}[E_1(\omega)] + \beta_2\mathrm{E}[E_2(\omega)] \tag{6.9}$$

where

$$\alpha_1 = c_1 \frac{\mathrm{E}[H_{12}(\omega)]\mathrm{E}[H_{22}(\omega)]}{D}, \quad \alpha_2 = c_2 \frac{\mathrm{E}[H_{21}(\omega)]\mathrm{E}[H_{11}(\omega)]}{D}, \tag{6.10,6.11}$$

$$\beta_1 = c_1 \frac{(\mathrm{E}[H_{22}(\omega)] - \mathrm{E}[H_{21}(\omega)] - \mathrm{E}[H_{dc_2}(\omega)])\mathrm{E}[H_{dc_1}(\omega)] - \mathrm{E}[H_{12}(\omega)]\mathrm{E}[H_{dc_2}(\omega)]}{D}, \tag{6.12}$$

$$\beta_2 = c_2 \frac{(\mathrm{E}[H_{11}(\omega)] - \mathrm{E}[H_{12}(\omega)] - \mathrm{E}[H_{dc_1}(\omega)])\mathrm{E}[H_{dc_2}(\omega)] - \mathrm{E}[H_{21}(\omega)]\mathrm{E}[H_{dc_1}(\omega)]}{D}, \tag{6.13}$$

and

$$D = (\mathrm{E}[H_{11}(\omega)] - \mathrm{E}[H_{dc_1}(\omega)])(\mathrm{E}[H_{22}(\omega)] - \mathrm{E}[H_{dc_2}(\omega)]) - \mathrm{E}[H_{12}(\omega)](\mathrm{E}[H_{dc_2}(\omega)])$$
$$-\mathrm{E}[H_{21}(\omega)](\mathrm{E}[H_{11}(\omega)] - \mathrm{E}[H_{dc_1}(\omega)]).$$

These equations are similar to those derived by Keane (1988). It is clear that the constants α_1, α_2, β_1 and β_2 are frequency dependent, since these equations deal with energy flows at a particular frequency. They also depend in a complicated way on both k_c and γ, and it is impossible to separate the expressions into distinct terms distinguishing the spring and damper dominant terms. They do, however, show that the energy flows are not simply related to the difference in the energy levels, even after taking ensemble averages.

7. Conclusions

Energy flow relationships for nonconservatively coupled rods have been established using a modal approach. It has been shown that the expressions derived for the various receptances are consistent with those for a conservative coupling when the stiffness is replaced by a complex coupling stiffness which includes the contributions of both spring and damper, but that additional terms proportional to the damper strength also arise (in fact, if the analysis leading to the standard results for conservatively coupled systems is carried out, but using a complex coupling stiffness, the extra terms are readily obtained). The effects of changes in the coupling parameters on the various power receptances have been illustrated through the use of numerical examples in which one rod is excited by 'rain-on-the-roof' forcing.

It has further been shown that the energy transferred to rod 2 through the coupling has a similar qualitative behaviour to that appertaining to the case of conservative coupling. Additionally, it is seen that the power dissipated within the coupling takes relatively low absolute levels except for a specific range of coupling damper rates. Outside of this range the damper is either so weak that it absorbs almost no power or is so strong that it virtually locks the two subsystems rigidly together. Moreover, within this range it is quite easy to arrange for almost no power to be transmitted through the coupling to the undriven subsystem, as might be expected.

Finally, a relationship between the average energy flows and the average total energies has been recovered. The results are consistent with those derived by Chen and Soong (1991) for two coupled oscillators although the constants of proportionality are, of course, different. They are seen to depend in a complicated way on the contributions from both the stiffness and the damping within the coupling and cannot be readily separated into distinct forms containing stiffness and damping dominant terms.

References

Chen, G. & Soong, T. T. 1991 Power flow and energy balance between nonconservatively coupled oscillators. *J. Sound Vib.* **113**, 448–454.

Crandal, S. H. & Lotz, R. 1971 On the coupling loss factor in statistical energy analysis. *J. Acoust. Soc. Am.* **49**, 352–356.

Davies, H. G. 1972*a* Exact solutions for the response of some coupled multimodal systems. *J. Acoust. Soc. Am.* **51**(1), 387–392.

Davies, H. G. 1972*b* Power flow between two coupled beams. *J. Acoust. Soc. Am.* **51**(1), 393–401.

Fahy, F. J. 1974 Statistical energy analysis – a critical review. *Shock and Vibration Digest* **6**, 14–33.

Fahy, F. J. & Mohammed, A. D. 1992 A study of uncertainty in applications of SEA to coupled beam and plate systems, Part I: Computational experiments. *J. Sound Vib.* **158**(1), 45–67.

Fahy, F. J. & Yao, D. 1987 Power flow between nonconservatively coupled oscillators. *J. Sound Vib.* **112**, 1–11

Goyden, H. G. D., White, R. G., Granch, E. T. & Adler, A. A. 1956 Bending vibrations of variable cross section beams. *Trans. of ASME J. Appl. Mech.* **23**(1), 103–108.

Keane, A. J. 1988 *Statistical Energy Analysis of Engineering Structures.* Ph.D. Thesis, Brunel University, Uxbridge.

Keane, A. J. & Price, W. G. 1987 Statistical energy analysis of strongly coupled systems. *J. Sound Vib.* **117**(2), 363–386.

Keane, A. J. & Price, W. G. 1991 A note on the power flowing between two conservatively coupled multi-modal sub-systems. *J. Sound Vib.* **144**(2), 185–196.

Lyon, R. H. 1975 *Statistical Energy Analysis of Dynamical Systems: Theory and Applications.* MIT Press.

Lyon, R. H. & Maidanik, G. 1962 Power flow between linearly coupled oscillators. *J. Acoust. Soc. Am.* **34**(5), 623–639.

Mace, B. R. 1993 The statistical energy analysis of two continuous one-dimensional subsystems. *J. Sound Vib.* **166**(3), 429–461

Meirovitch, L. 1975 *Elements of Vibration Analysis.* McGraw-Hill.

Newland, D. E. 1975 *An Introduction to Random Vibrations and Spectral Analysis (2nd edition).* Longman.

Norton, M. 1989 *Fundamentals of Noise and Vibration Analysis for Engineers.* Cambridge University Press.

Remington, P. J. & Manning, J. E. 1975 Comparison of statistical energy analysis power flow predictions with and 'exact' calculation. *J. Acoust. Soc. Am.* **57**(2), 374–379.

Scharton, T. D. & Lyon, R. H. 1968 Power flow and energy sharing in random vibration. *J. Acoust. Soc. Am.* **43**(6), 1332–1343.

Sun, J. C., Lalor, N. & Richards, E. J. 1987 Power flow and energy balance of nonconservatively coupled structures, 1: Theory. *J. Sound Vib.* **112**, 321–330.